Excel 2010 for Physical Sciences Statistics

Thomas J. Quirk • Meghan Quirk
Howard Horton

Excel 2010 for Physical Sciences Statistics

A Guide to Solving Practical Problems

 Springer

Thomas J. Quirk
Professor of Marketing
Webster University
St. Louis, MO, USA

Meghan Quirk
Bailey, CO, USA

Howard Horton
Bailey, CO, USA

ISBN 978-3-319-00629-1   ISBN 978-3-319-00630-7 (ebook)
DOI 10.1007/978-3-319-00630-7
Springer Cham Heidelberg New York Dordrecht London

Library of Congress Control Number: 2013941475

Printed on acid-free paper

Springer is part of Springer Science+Business Media (www.springer.com)

This book is dedicated to the more than 3,000 students I have taught at Webster University's campuses in St. Louis, London, and Vienna; the students at Principia College in Elsah, Illinois; and the students at the Cooperative State University of Baden-Wuerttemburg in Heidenheim, Germany. These students taught me a great deal about the art of teaching. I salute them all, and I thank them for helping me to become a better teacher.

Thomas J. Quirk

We dedicate this book to all the newly-inspired students emerging into the ranks of the various fields of science.

Meghan Quirk and Howard Horton

Preface

Excel 2010 for Physical Sciences Statistics: A Guide to Solving Practical Problems is intended for anyone looking to learn the basics of applying Excel's powerful statistical tools to their science courses or work activities. If understanding statistics isn't your strongest suit, you are not especially mathematically-inclined, or if you are wary of computers, then this is the right book for you.

Here you'll learn how to use key statistical tests using Excel without being overpowered by the underlying statistical theory. This book clearly and methodically shows and explains how to create and use these statistical tests to solve practical problems in the physical sciences.

Excel is an easily available computer program for students, instructors, and managers. It is also an effective teaching and learning tool for quantitative analyses in science courses. The powerful numerical computational ability and the graphical functions available in Excel make learning statistics much easier than in years past. However, this is the first book to show Excel's capabilities to more effectively teach science statistics; it also focuses exclusively on this topic in an effort to render the subject matter not only applicable and practical, but also easy to comprehend and apply.

Unique features of this book:

- This book is appropriate for use in any course in the Physical Sciences Statistics (at both undergraduate and graduate levels) as well as for managers who want to improve the usefulness of their Excel skills.
- Includes 159 color screen shots so that you can be sure you are performing the Excel steps correctly
- You will be told each step of the way, not only *how* to use Excel, but also *why* you are doing each step so that you can understand what you are doing, and not merely learn how to use statistical tests by rote.
- Includes specific objectives embedded in the text for each concept, so you can know the purpose of the Excel steps.
- This book is a tool that can be used either by itself or along with *any* good statistics book.

- Statistical theory and formulas are explained in clear language without bogging you down in mathematical fine points.
- You will learn both how to write statistical formulas using Excel and how to use Excel's drop-down menus that will create the formulas for you.
- This book does not come with a CD of Excel files which you can upload to your computer. Instead, you'll be shown how to create each Excel file yourself. In a work situation, your colleagues will not give you an Excel file; you will be expected to create your own. This book will give you ample practice in developing this important skill.
- Each chapter presents the steps needed to solve a practical science problem using Excel. In addition, there are three practice problems at the end of each chapter so you can test your new knowledge of statistics. The answers to these problems appear in Appendix A.
- A "Practice Test" is given in Appendix B to test your knowledge at the end of the book. The answers to these practical science problems appear in Appendix C.

Thomas Quirk, a current Professor of Marketing at the George Herbert Walker School of Business & Technology at Webster University in St. Louis, Missouri (USA), teaches Marketing Statistics, Marketing Research, and Pricing Strategies. He has published articles in *The Journal of Educational Psychology, Journal of Educational Research, Review of Educational Research, Journal of Educational Measurement, Educational Technology, The Elementary School Journal, Journal of Secondary Education, Educational Horizons, and Phi Delta Kappan*. In addition, Professor Quirk has written more than 60 textbook supplements in Management and Marketing, published more than 20 articles in professional journals, and presented more than 20 papers at professional meetings. He holds a BS in Mathematics from John Carroll University, both a MA in Education and a PhD in Educational Psychology from Stanford University, and an MBA from the University of Missouri-St. Louis.

Meghan H. Quirk holds both a PhD in Biological Education and an MA in Biological Sciences from the University of Northern Colorado (UNC), and a BA in Biology and Religion at Principia College in Elsah, Illinois. She has done research on foodweb dynamics at Wind Cave National Park in South Dakota and research in agro-ecology in Southern Belize. She has co-authored an article on shortgrass steppe ecosystems in *Photochemistry & Photobiology* and has presented papers at the Shortgrass Steppe Symposium in Fort Collins, Colorado, the Long-term Ecological Research All Scientists Meeting in Estes Park, Colorado, and participated in the NSF Site Review of the Shortgrass Steppe Long Term Ecological Research in Nunn, Colorado. She was a National Science Foundation Fellow GK-12, and currently teaches in Bailey, Colorado.

Howard F. Horton holds an MS in Biological Sciences from the University of Northern Colorado (UNC) and a BS in Biological Sciences from Mesa State College. He has worked on research projects in Pawnee National Grasslands, Rocky Mountain National Park, Long Term Ecological Research at Toolik Lake, Alaska, and Wind Cave, South Dakota. He has co-authored articles in *The International*

Journal of Speleology and *The Journal of Cave and Karst Studies*. He was a National Science Foundation Fellow GK-12, and has worked as a District Wildlife Manager with the Colorado Division of Parks and Wildlife. He is currently the Angler Outreach Coordinator with Colorado Parks and Wildlife in the USA.

MO, USA   Thomas J. Quirk
CO, USA   Meghan Quirk
CO, USA   Howard Horton

Acknowledgements

Excel 2010 for Physical Sciences Statistics: A Guide to Solving Practical Problems is the result of inspiration from three important people: my two daughters and my wife. Jennifer Quirk McLaughlin invited me to visit her MBA classes several times at the University of Witwatersrand in Johannesburg, South Africa. These visits to a first-rate MBA program convinced me there was a need for a book to teach students how to solve practical problems using Excel. Meghan Quirk-Horton's dogged dedication to learning the many statistical techniques needed to complete her PhD dissertation illustrated the need for a statistics book that would make this daunting task more user-friendly. And Lynne Buckley-Quirk was the number-one cheerleader for this project from the beginning, always encouraging me and helping me remain dedicated to completing it.

Thomas J. Quirk

We would like to acknowledge the patience of our two little girls, Lila and Elia, as we worked on this book with their TQ. We would also like to thank Professors Sarah Perkins, Doug Warren, John Moore, and Lee Dyer for their guidance and support during our college and graduate school careers.

Meghan Quirk and Howard Horton

Marc Strauss, our editor at Springer, caught the spirit of this idea in our first phone conversation and shepherded this book through the idea stages until it reached its final form. His encouragement and support were vital to this book seeing the light of day. We thank him for being such an outstanding product champion throughout this process.

Contents

Chapter 1
Sample Size, Mean, Standard Deviation, and Standard Error of the Mean

This chapter deals with how you can use Excel to find the average (i.e., "mean") of a set of scores, the standard deviation of these scores (STDEV), and the standard error of the mean (s.e.) of these scores. All three of these statistics are used frequently and form the basis for additional statistical tests.

1.1 Mean

The *mean* is the "arithmetic average" of a set of scores. When my daughter was in the fifth grade, she came home from school with a sad face and said that she didn't get "averages." The book she was using described how to find the mean of a set of scores, and so I said to her:

> "Jennifer, you add up all the scores and divide by the number of numbers that you have."
> She gave me "that look," and said: "Dad, this is serious!" She thought I was teasing her.
> So I said:
> "See these numbers in your book; add them up. What is the answer?" (She did that.)
> "Now, how many numbers do you have?" (She answered that question.)
> "Then, take the number you got when you added up the numbers, and divide that number by the number of numbers that you have."

She did that, and found the correct answer. You will use that same reasoning now, but it will be much easier for you because Excel will do all of the steps for you.

We will call this average of the scores the "mean" which we will symbolize as: $\overline{X}$, and we will pronounce it as: "Xbar."

The formula for finding the mean with you calculator looks like this:

$$\overline{X} = \frac{\Sigma X}{n} \tag{1.1}$$

T.J. Quirk et al., *Excel 2010 for Physical Sciences Statistics: A Guide to Solving Practical Problems*, DOI 10.1007/978-3-319-00630-7_1,
© Springer International Publishing Switzerland 2013

The symbol Σ is the Greek letter sigma, which stands for "sum". It tells you to add up all the scores that are indicated by the letter X, and then to divide your answer by n (the number of numbers that you have).

Let's give a simple example:

Suppose that you had these six chemistry test scores on an 7-item true-false quiz:

6
4
5
3
2
5

To find the mean of these scores, you add them up, and then divide by the number of scores. So, the mean is: 25/6 = 4.17

1.2 Standard Deviation

The *standard deviation* tells you "how close the scores are to the mean." If the standard deviation is a small number, this tells you that the scores are "bunched together" close to the mean. If the standard deviation is a large number, this tells you that the scores are "spread out" a greater distance from the mean. The formula for the standard deviation (which we will call STDEV) and use the letter, S, to symbolize is:

$$\text{STDEV} = S = \sqrt{\frac{\Sigma(X - \overline{X})^2}{n - 1}} \qquad (1.2)$$

The formula look complicated, but what it asks you to do is this:

1. Subtract the mean from each score $(X - \overline{X})$.
2. Then, square the resulting number to make it a positive number.
3. Then, add up these squared numbers to get a total score.
4. Then, take this total score and divide it by n − 1 (where n stands for the number of numbers that you have).
5. The final step is to take the square root of the number you found in step 4.

You will not be asked to compute the standard deviation using your calculator in this book, but you could see examples of how it is computed in any basic statistics book (e.g. Schuenemeyer and Drew 2011). Instead, we will use Excel to find the standard deviation of a set of scores. When we use Excel on the six numbers we gave in the description of the mean above, you will find that the *STDEV* of these numbers, S, is 1.47.

1.3   Standard Error of the Mean

The formula for the *standard error of the mean* (*s.e.*, which we will use $S_{\overline{X}}$ to symbolize) is:

$$s.e. = S_{\overline{X}} = \frac{S}{\sqrt{n}} \tag{1.3}$$

To find **s.e.**, all you need to do is to take the standard deviation, STDEV, and divide it by the square root of n, where n stands for the "number of numbers" that you have in your data set. In the example under the standard deviation description above, the *s.e.* = 0.60 . (You can check this on your calculator).

If you want to learn more about the standard deviation and the standard error of the mean, see McKillup and Dyar (2010) and Schuenemeyer and Drew (2011).

Now, let's learn how to use Excel to find the sample size, the mean, the standard deviation, and the standard error or the mean using the level of sulphur dioxide in rainfall measured in milligrams (mg) of sulphur per liter (L) of rainfall. (Note that one milligram (mg) equals one thousandth of one gram and is a metric measure of weight, while one liter is a metric unit of the volume of one kilogram of pure water under standard conditions.) Suppose that eight samples of rainfall were taken. The hypothetical data appear in Fig. 1.1.

Fig. 1.1 Worksheet Data for Sulphur Dioxide Levels (Practical Example)

Sample	milligrams per liter (mg/L)
1	0.4
2	1.2
3	0.8
4	0.4
5	1.3
6	0.6
7	0.7
8	1.1

1.4 Sample Size, Mean, Standard Deviation, and Standard Error of the Mean

> Objective: To find the sample size (n), mean, standard deviation (STDEV), and standard error of the mean (s.e.) for these data

Start your computer, and click on the Excel 2010 icon to open a blank Excel spreadsheet.

Enter the data in this way:

B3:   Sample
C3:   milligrams per liter (mg/L)
B4:   1

1.4.1 Using the Fill/Series/Columns Commands

> Objective: To add the sample numbers 2–8 in a column underneath Sample #1

Put pointer in B4
Home (top left of screen)
Fill (top right of screen: click on the down arrow; see Fig. 1.2)

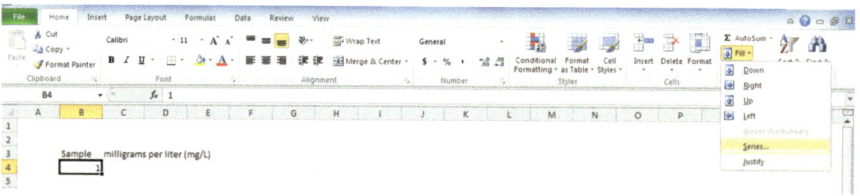

Fig. 1.2 Home/Fill/Series commands

Series
Columns
Step value: 1
Stop value: 8 (see Fig. 1.3)

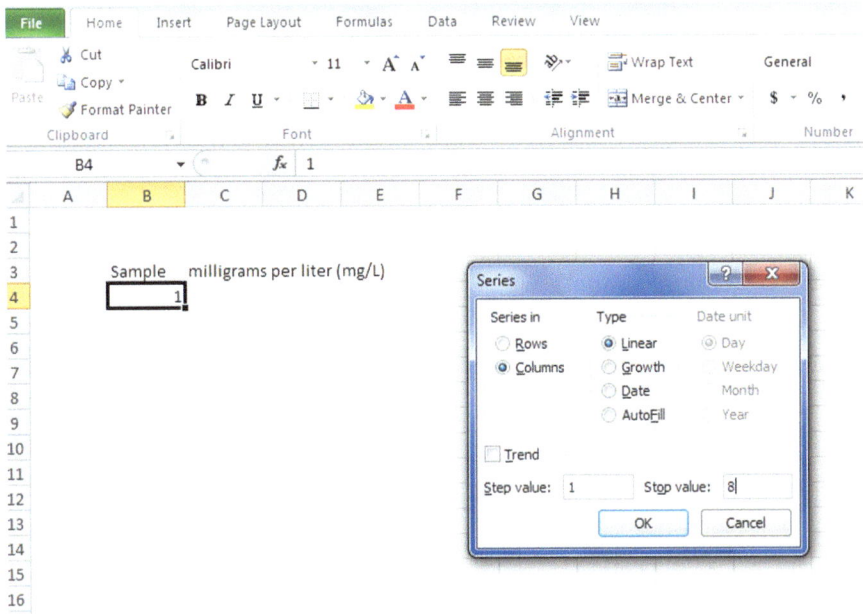

Fig. 1.3 Example of Dialogue Box for Fill/Series/Columns/Step Value/Stop Value commands

OK

The sample numbers should be identified as 1–8, with 8 in cell B11.

Now, enter the milligrams per liter in cells C4: C11. *(Note: Be sure to double-check your figures to make sure that they are correct or you will not get the correct answer!)*

Since your computer screen shows the information in a format that does not look professional, you need to learn how to "widen the column width" and how to "center the information" in a group of cells. Here is how you can do those two steps:

1.4.2 Changing the Width of a Column

Objective: To make a column width wider so that all of the information fits inside that column

If you look at your computer screen, you can see that Column C is not wide enough so that all of the information fits inside this column. To make Column C wider:

Click on the letter, C, at the top of your computer screen

Place your mouse pointer on your computer at the far right corner of C until you createa "cross sign" on that corner

Left-click on your mouse, hold it down, and move this corner to the right until it is "wide enough to fit all of the data"

Take your finger off your mouse to set the new column width (see Fig. 1.4)

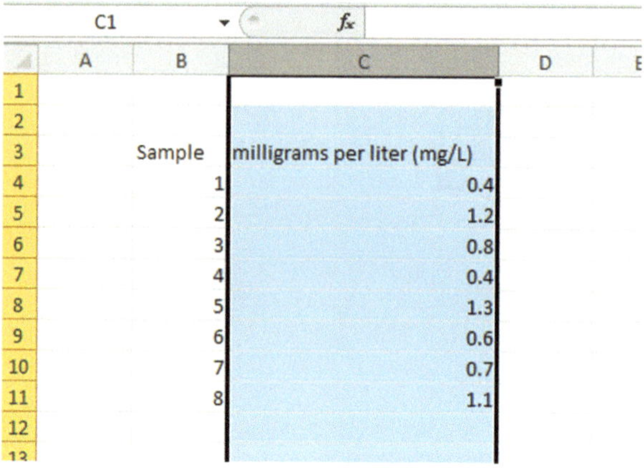

Fig. 1.4 Example of How to Widen the Column Width

Then, click on any empty cell (i.e., any blank cell) to "deselect" column C so that it is no longer a darker color on your screen.

When you widen a column, you will make all of the cells in all of the rows of this column that same width.

Now, let's go through the steps to center the information in both Column B and Column C.

1.4.3   Centering Information in a Range of Cells

Objective: To center the information in a group of cells

In order to make the information in the cells look "more professional," you can center the information using the following steps:

Left-click your mouse pointer on B3 and drag it to the right and down to highlight cells B3:C11 so that these cells appear in a darker color

At the top of your computer screen, you will see a set of "lines" in which all of the lines are "centered" to the same width under "Alignment" (it is the second icon at the bottom left of the Alignment box; see Fig. 1.5)

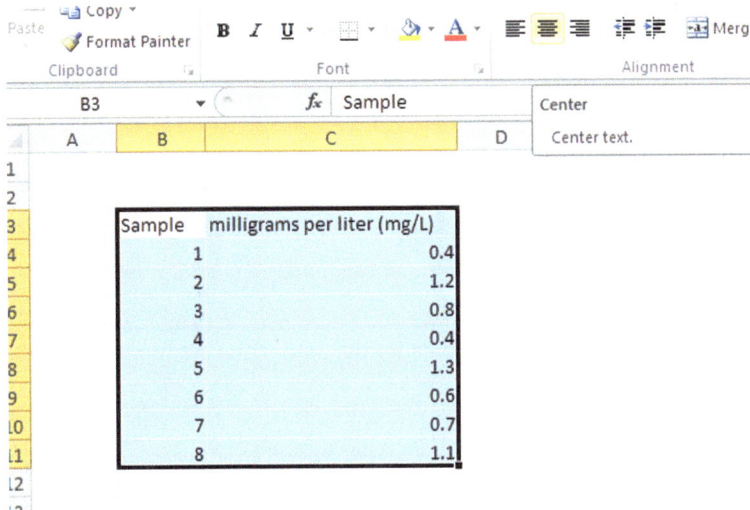

Fig. 1.5 Example of How to Center Information Within Cells

Click on this icon to center the information in the selected cells (see Fig. 1.6)

	A	B	C	D
1				
2				
3		Sample	milligrams per liter (mg/L)	
4		1	0.4	
5		2	1.2	
6		3	0.8	
7		4	0.4	
8		5	1.3	
9		6	0.6	
10		7	0.7	
11		8	1.1	
12				
13				

Fig. 1.6 Final Result of Centering Information in the Cells

Since you will need to refer to the milligrams per liter in your formulas, it will be much easier to do this if you "name the range of data" with a name instead of having to remember the exact cells (C4 : C11) in which these figures are located. Let's call that group of cells:Weight, but we could give them any name that you want to use.

1.4.4   Naming a Range of Cells

Objective: To name the range of data for the milligrams per liter with the name:
          Weight

Highlight cells C4 : C11 by left-clicking your mouse pointer on C4 and dragging it
    down to C11
Formulas (top left of your screen)
Define Name (top center of your screen)
Weight (type this name in the top box; see Fig. 1.7)

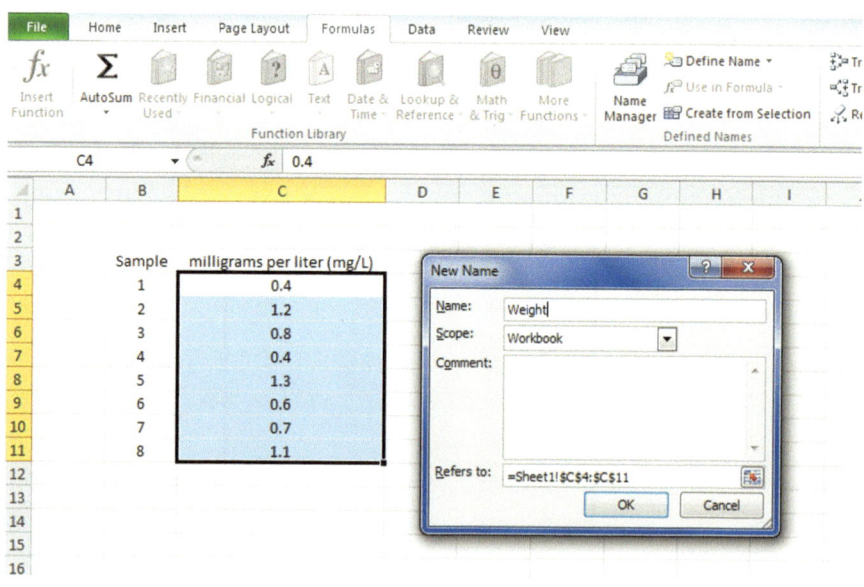

Fig. 1.7 Dialogue box for "naming a range of cells" with the name: Weight

OK

Then, click on any cell of your spreadsheet that doesnot have any information in it
(i.e., it is an"empty cell") to deselect cells C4:C11
Now, add the following terms to your spreadsheet:

E6:   n
E9:   Mean
E12:   STDEV
E15:   s.e. (see Fig. 1.8)

	A	B	C	D	E	F
1						
2						
3		Sample	milligrams per liter (mg/L)			
4		1	0.4			
5		2	1.2			
6		3	0.8		n	
7		4	0.4			
8		5	1.3			
9		6	0.6		Mean	
10		7	0.7			
11		8	1.1			
12					STDEV	
13						
14						
15					s.e.	
16						
17						

Fig. 1.8   Example of Entering the Sample Size, Mean, STDEV, and s.e. Labels

*Note: Whenever you use a formula, you must add an equal sign (=) at the beginning of
the name of the function so that Excel knows that you intend to use a formula.*

1.4.5   Finding the Sample Size Using the =COUNT Function

Objective: To find the sample size (n) for these data using the =COUNT
function

F6:   =COUNT(Weight)

This command should insert the number 8 into cell F6 since there are eight samples of rainfall in your sample.

1.4.6 Finding the Mean Score Using the =AVERAGE Function

> Objective: To find the mean weight figure using the =AVERAGE function

F9:   =AVERAGE(Weight)

This command should insert the number 0.8125 into cell F9.

1.4.7 Finding the Standard Deviation Using the =STDEV Function

> Objective: To find the standard deviation (STDEV) using the =STDEV function

F12:   =STDEV(Weight)

This command should insert the number 0.352288 into cell F12.

1.4.8 Finding the Standard Error of the Mean

> Objective: To find the standard error of the mean using a formula for these eight data points

F15:   =F12/SQRT(8)

This command should insert the number 0.124553 into cell F15 (see Fig. 1.9).

	A	B	C	D	E	F	G
1							
2							
3		Sample	milligrams per liter (mg/L)				
4		1	0.4				
5		2	1.2				
6		3	0.8		n	8	
7		4	0.4				
8		5	1.3				
9		6	0.6		Mean	0.8125	
10		7	0.7				
11		8	1.1				
12					STDEV	0.352288	
13							
14							
15					s.e.	0.124553	
16							
17							

Fig. 1.9 Example of Using Excel Formulas for Sample Size, Mean, STDEV, and s.e.

Important note: Throughout this book, be sure to double-check all of the figures in your spreadsheet to make sure that they are in the correct cells, or the formulas will not work correctly!

1.4.8.1   Formatting Numbers in Number Format (2 decimal places)

Objective: To convert the mean, STDEV, and s.e. to two decimal places

Highlight cells F9 : F15
Home (top left of screen)
Look under "Number" at the top center of your screen. In the bottom right corner, gently place your mouse pointer on you screen at the bottom of the .00 .0 until it says: "Decrease Decimals" (see Fig. 1.10)

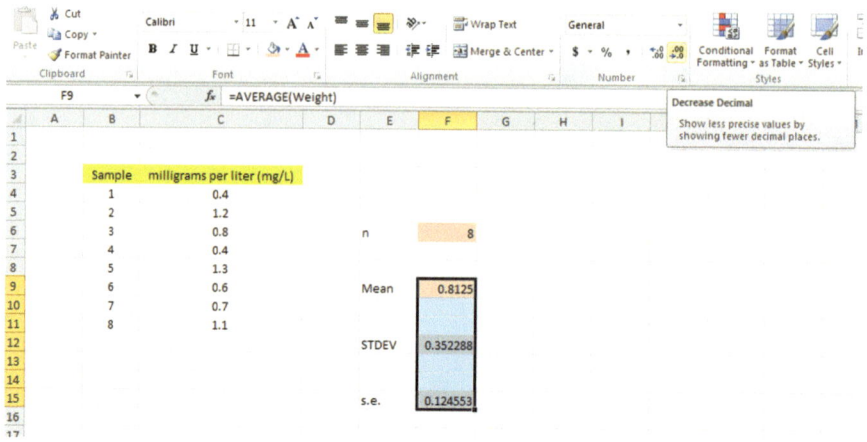

Fig. 1.10 Using the "Decrease Decimal Icon" to convert Numbers to Fewer Decimal Places

Click on this icon *twice* and notice that the cells F9:F15 are now all in just two decimal places (see Fig. 1.11)

	A	B	C	D	E	F	G
1							
2							
3		Sample	milligrams per liter (mg/L)				
4		1	0.4				
5		2	1.2				
6		3	0.8		n	8	
7		4	0.4				
8		5	1.3				
9		6	0.6		Mean	0.81	
10		7	0.7				
11		8	1.1				
12					STDEV	0.35	
13							
14							
15					s.e.	0.12	
16							
17							

Fig. 1.11 Example of Converting Numbers to Two Decimal Places

Now, click on any "empty cell" on your spreadsheet to deselect cells F9:F15.

1.5   Saving a Spreadsheet

Objective: To save this spreadsheet with the name: sulphur3

In order to save your spreadsheet so that you can retrieve it sometime in the future, your first decision is to decide "where" you want to save it. That is your decision and you have several choices. If it is your own computer, you can save it onto your hard drive (you need to ask someone how to do that on your computer). Or, you can save it onto a "CD" or onto a "flash drive." You then need to complete these steps:

File
Save as

(select the place where you want to save the file by scrolling either down or up the bar on the left, and click on the place where you want to save the file; for example: Documents: My Documents location)

File name: sulphur3 (enter this name to the right of File name; see Fig. 1.12)
Save

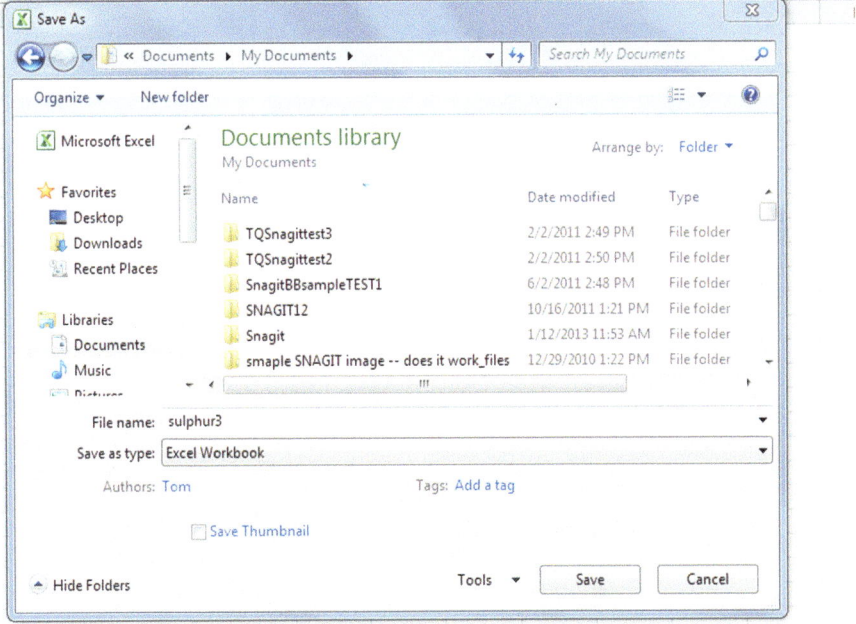

Fig. 1.12 Dialogue Box of Saving an Excel Workbook File as "sulphur3" in Documents: My Documents location

Important note: Be very careful to save your Excel file spreadsheet every few minutes so that you do not lose your information!

1.6   Printing a Spreadsheet

> Objective: To print the spreadsheet

Use the following procedure when printing any spreadsheet.

File
Print
Print Active Sheets (see Fig. 1.13)
Print (top of your screen)

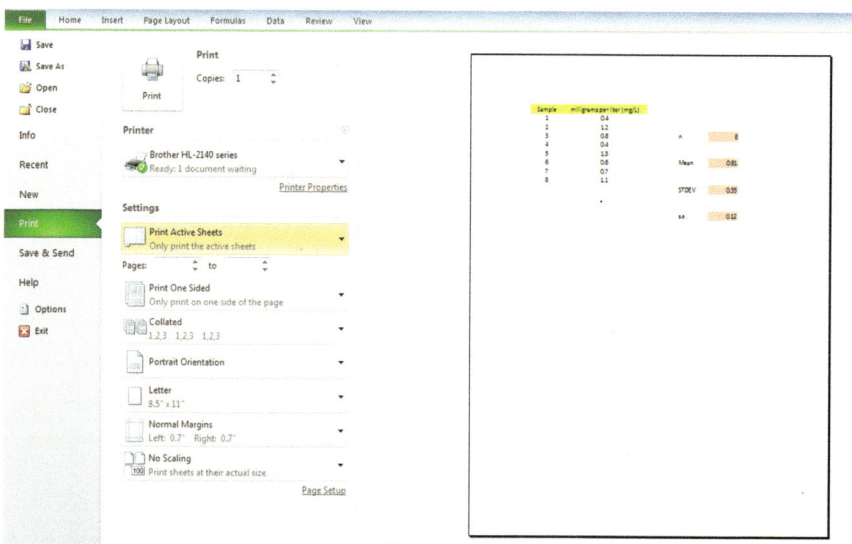

Fig. 1.13 Example of How to Print an Excel Worksheet Using the File/Print/Print Active Sheets Commands

The final spreadsheet is given in Fig. 1.14

◢	A	B	C	D	E	F	G
1							
2							
3		Sample	milligrams per liter (mg/L)				
4		1	0.4				
5		2	1.2				
6		3	0.8		n	8	
7		4	0.4				
8		5	1.3				
9		6	0.6		Mean	0.81	
10		7	0.7				
11		8	1.1				
12					STDEV	0.35	
13							
14							
15					s.e.	0.12	
16							
17							

Fig. 1.14   Final Result of Printing an Excel Spreadsheet

Before you leave this chapter, let's practice changing the format of the figures on a spreadsheet with two examples: (1) using two decimal places for figures that are dollar amounts, and (2) using three decimal places for figures.

Close your spreadsheet by: File/Close/Don't Save, and open a blank Excel spreadsheet by usingFile/New/Create (on the far right of your screen).

1.7   Formatting Numbers in Currency Format (2 decimal places)

Objective: To change the format of figures to dollar format with two decimal
places

A3:   Price
A4:   1.25
A5:   3.45
A6:   12.95

Home
Highlight cells A4:A6 by left-clicking your mouse on A4 and dragging it down so that these three cells are highlighted in a darker color
Number (top center of screen: click on the down arrow on the right; see Fig. 1.15)

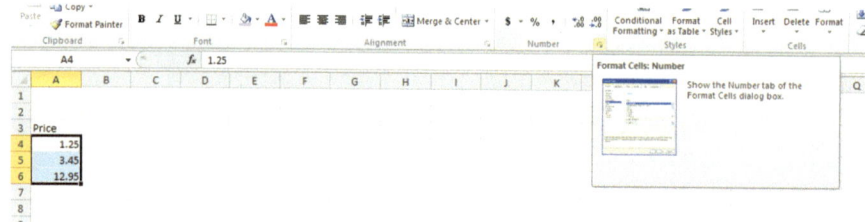

Fig. 1.15 Dialogue Box for Number Format Choices

Category: Currency
Decimal places: 2 (then see Fig. 1.16)

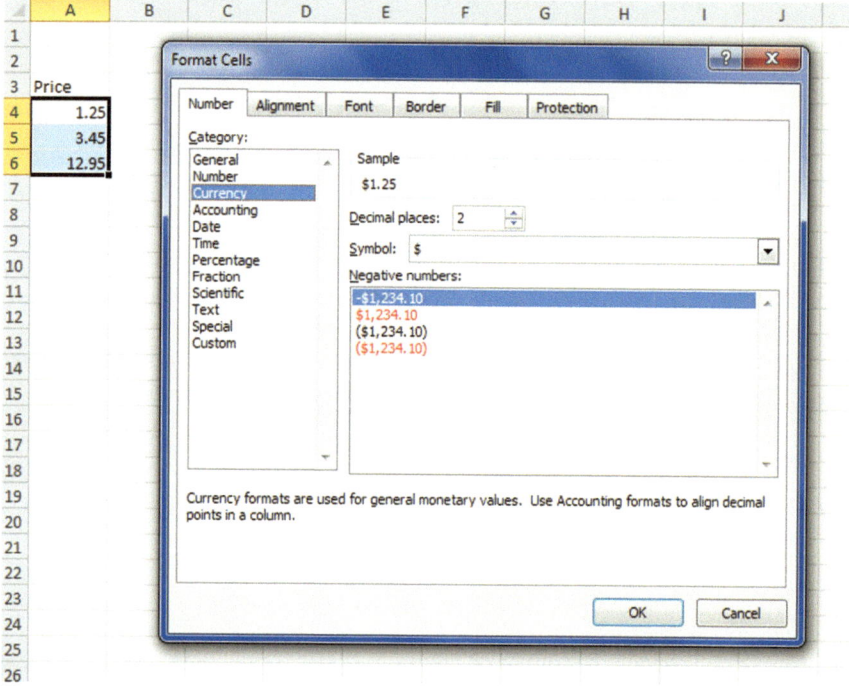

Fig. 1.16 Dialogue Box for Currency (2 decimal places) Format for Numbers

OK

The three cells should have a dollar sign in them and be in two decimal places.
Next, let's practice formatting figures in number format, three decimal places.

1.8   Formatting Numbers in Number Format (3 decimal places)

Objective: To format figures in number format, three decimal places

Home
Highlight cells A4:A6 on your computer screen
Number (click on the down arrow on the right)
Category: number
At the right of the box, change 2 decimal places to 3 decimal places by clicking on the "up arrow" once
OK

The three figures should now be in number format, each with three decimals.
Now, click on any blank cell to deselect cells A4:A6. Then, close this file by File/Close/Don't Save (since there is no need to save this practice problem).
You can use these same commands to format a range of cells in percentage format (and many other formats) to whatever number of decimal places you want to specify.

1.9   End-of-Chapter Practice Problems

1. Limonite is a type of mineral that includes many other minerals. It forms the color of many soils and on the weathered surfaces of rocks. It also exists in iron ore. Suppose that you wanted to find the mean, standard deviation, and standard error of the mean for the percent of iron in limonite (iron ore) samples. The hypothetical data appear in Fig. 1.17.

Fig. 1.17 Worksheet Data for Chapter 1: Practice Problem #1

IRON ORE (LIMONITE) SAMPLES

Percent (%) iron
18.24
18.29
18.26
18.28
18.30
18.24
18.26
18.25
18.28
18.29
18.30
18.24
18.26

(a) Use Excel to the right of the table to find the sample size, mean, standard deviation, and standard error of the mean for these data. Label your answers, and round off the mean, standard deviation, and standard error of the mean to two decimal places; use number format for these three figures.

(b) Print the result on a separate page.

(c) Save the file as: iron3

2. Suppose that you have been hired as a research assistant and that you have been asked to determine the average micrograms of lead concentration per cubic meter ($\mu g/m^3$) for air samples taken near Route 101 near San Francisco in weekday afternoons between 4 p.m. and 7 p.m. The hypothetical data are given in Figure 1.18.

LEAD CONCENTRATION IN AIR SAMPLES TAKEN NEAR SAN FRANCISCO

Micrograms per cubic meter ($\mu g/m^3$)
3.1
10.1
6.7
8.9
5.6
6.4
4.8
10.2
9.8
8.4
7.5
9.4
8.5
4.8

Fig. 1.18 Worksheet Data for Chapter 1: Practice Problem #2

(a) Use Excel to create a table of these data, and at the right of the table use Excel to find the sample size, mean, standard deviation, and standard error of the mean for these data. Label your answers, and round off the mean, standard deviation, and standard error of the mean to two decimal places using number format.

(b) Print the result on a separate page.

(c) Save the file as: air3

3. Suppose that you have been asked to measure the percent of silver found in various silver ore samples from a mine. The mine has provided 16 ore samples from different locations within the mine. You have processed the ore samples to determine the amount of silver in each sample. The hypothetical data are given in Fig. 1.19:

Fig. 1.19 Worksheet Data for Chapter 1: Practice Problem #3

SILVER ORE SAMPLES

PERCENT (%) SILVER
12
15
13
8
10
12
13
12
9
4
11
15
13
15
12
14

(a) Use Excel to create a table for these data, and at the right of the table, use Excel to find the sample size, mean, standard deviation, and standard error of the mean for these data. Label your answers, and round off the mean, standard deviation, and standard error of the mean to three decimal places using number format.

(b) Print the result on a separate page.

(c) Save the file as: SILVER3

References

McKillup S., Dyar M. Geostatistics Explained: an introductory guide for earth scientists. Cambridge: Cambridge University Press; 2010.

Schuenemeyer J, Drew L. Statistics for Earth and Environmental Scientists. Hoboken: John Wiley & Sons; 2011.

Chapter 2
Random Number Generator

Salt marshes are coastal wetlands found on protected shorelines along the eastern seaboard of the USA where fresh water mixes with seawater. When ocean tides flood salt marshes, the plants living there must cope with the salt water. The "salinity" (i.e., the salt content of the water) depends on how close the marsh is to the ocean. Suppose that a biogeographer is studying the effects of salinity on vegetation in a salt marsh in Maine and that she has mapped the salt marsh into 32 separate geographic areas. Suppose, further, that she has asked you to take a random sample of 5 of these 32 areas within the salt marsh so that she can measure the percent of salinity level in each of these areas. Using your Excel skills to take this random sample, you will need to define a "sampling frame."

A sampling frame is a list of objects, events, or people from which you want to select a random sample. In this case, it is the group of 32 areas of the salt marsh. The frame starts with the identification code (ID) of the number 1 that is assigned to the first area in the group of 32 areas. The second area has a code number of 2, the third a code number of 3, and so forth until the last area has a code number of 32.

Since the salt marsh has 32 areas, your sampling frame would go from 1 to 32 with each area having a unique ID number.

We will first create the frame numbers as follows in a new Excel worksheet:

2.1 Creating Frame Numbers for Generating Random Numbers

Objective: To create the frame numbers for generating random numbers

A3: FRAME NO.
A4: 1

T.J. Quirk et al., *Excel 2010 for Physical Sciences Statistics: A Guide to Solving Practical Problems*, DOI 10.1007/978-3-319-00630-7_2,
© Springer International Publishing Switzerland 2013

Now, create the frame numbers in column A with the Home/Fill commands that were explained in the first chapter of this book (see Sect. 1.4.1) so that the frame numbers go from 1 to 32, with the number 32 in cell A35. If you need to be reminded about how to do that, here are the steps:

Click on cell A4 to select this cell
Home
Fill (then click on the "down arrow" next to this command and select)
Series (see Fig. 2.1)

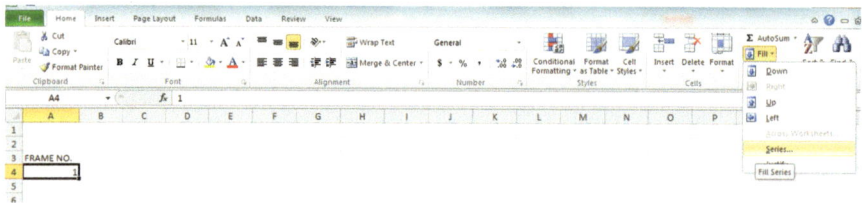

Fig. 2.1 Dialogue Box for Fill/Series Commands

Columns
Step value: 1
Stop value: 32 (see Fig. 2.2)

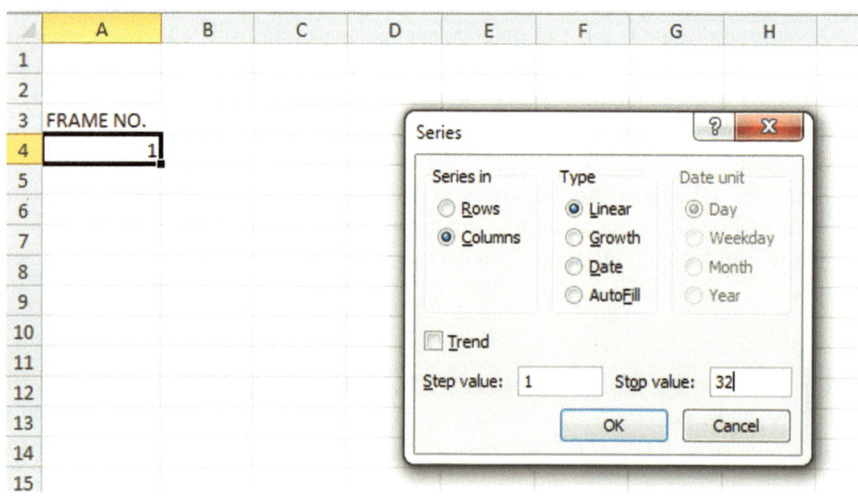

Fig. 2.2 Dialogue Box for Fill/Series/Columns/Step value/ Stop value Commands

OK
Then, save this file as: Random29. You should obtain the result in Fig. 2.3.

Fig. 2.3 Frame Numbers
from 1 to 32

FRAME NO.

1
2
3
4
5
6
7
8
9
10
11
12
13
14
15
16
17
18
19
20
21
22
23
24
25
26
27
28
29
30
31
32

Now, create a column next to these frame numbers in this manner:

B3: DUPLICATE FRAME NO.
B4:   1

Next, use the Home/Fill command again, so that the 32 frame numbers begin in cell B4 and end in cell B35. Be sure to widen the columns A and B so that all of the information in these columns fits inside the column width. Then, center the information inside both Column A and Column B on your spreadsheet. You should obtain the information given in Fig. 2.4.

Fig. 2.4 Duplicate Frame
Numbers from 1 to 32

FRAME NO.	DUPLICATE FRAME NO.
1	1
2	2
3	3
4	4
5	5
6	6
7	7
8	8
9	9
10	10
11	11
12	12
13	13
14	14
15	15
16	16
17	17
18	18
19	19
20	20
21	21
22	22
23	23
24	24
25	25
26	26
27	27
28	28
29	29
30	30
31	31
32	32

Save this file as: Random30

You are probably wondering why you created the same information in both Column A and Column B of your spreadsheet. This is to make sure that before you sort the frame numbers that you have exactly 32 of them when you finish sorting them into a random sequence of 32 numbers.

Now, let's add a random number to each of the duplicate frame numbers as follows:

2.2   Creating Random Numbers in an Excel Worksheet

C3:   RANDOM NO. (then widen columns A, B, C so that their labels fit inside the columns; then center the information in A3:C35)

C4:   =RAND()

Next, hit the Enter key to add a random number to cell C4.

Note that you need *both* an open parenthesis *and* a closed parenthesis after =*RAND*(). The RAND command "looks to the left of the cell with the RAND() COMMAND in it" and assigns a random number to that cell.

Now, put the pointer using your mouse in cell C4 and then move the pointer to the bottom right corner of that cell until you see a "plus sign" in that cell. Then, click and drag the pointer down to cell C35 to add a random number to all 32 ID frame numbers (see Fig. 2.5).

Fig. 2.5 Example of
Random Numbers Assigned
to the Duplicate Frame
Numbers

FRAME NO.	DUPLICATE FRAME NO.	RANDOM NO.
1	1	0.178997426
2	2	0.269196787
3	3	0.48649709
4	4	0.882904516
5	5	0.015953504
6	6	0.099651545
7	7	0.42850057
8	8	0.381659988
9	9	0.431296832
10	10	0.476642453
11	11	0.268603728
12	12	0.871330234
13	13	0.775421903
14	14	0.908450998
15	15	0.138749452
16	16	0.159535582
17	17	0.672417279
18	18	0.956231064
19	19	0.486746795
20	20	0.83596565
21	21	0.688574546
22	22	0.467838617
23	23	0.695493167
24	24	0.226521237
25	25	0.335451708
26	26	0.209245145
27	27	0.631291464
28	28	0.210229448
29	29	0.553196562
30	30	0.494647331
31	31	0.986702143
32	32	0.178067956

Then, click on any empty cell to deselect C4:C35 to remove the dark color highlighting these cells.

Save this file as: Random31

Now, let's sort these duplicate frame numbers into a random sequence:

2.3   Sorting Frame Numbers into a Random Sequence

Objective: To sort the duplicate frame numbers into a random sequence

Highlight cells B3 : C35 (include the labels at the top of columns B and C)
Data (top of screen)
Sort (click on this word at the top center of your screen; see Fig. 2.6)

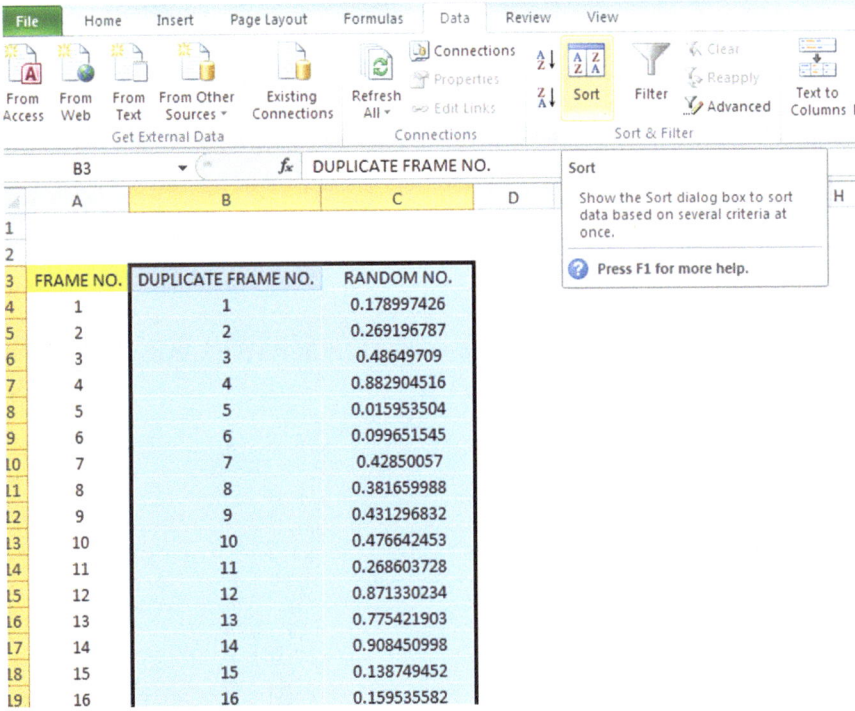

Fig. 2.6 Dialogue Box for Data/Sort Commands

Sort by: RANDOM NO. (click on the down arrow)
Smallest to Largest (see Fig. 2.7)

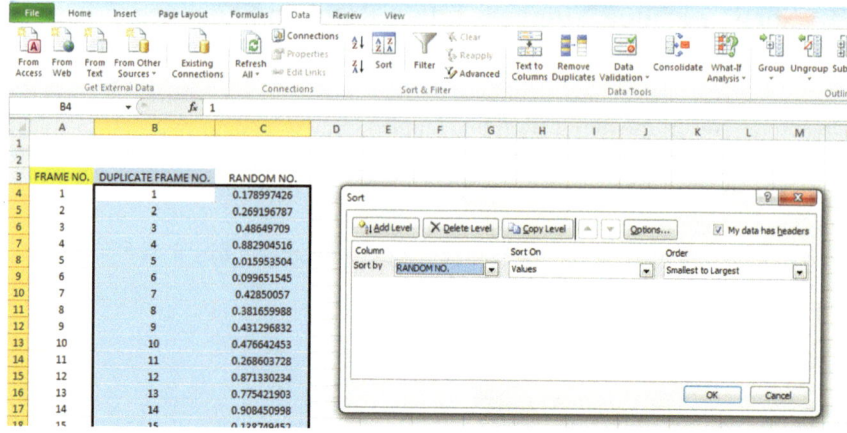

Fig. 2.7 Dialogue Box for Data/Sort/RANDOM NO./Smallest to Largest Commands

OK

Click on any empty cell to deselect B3:C35.

Save this file as: Random32

Print this file now.

These steps will produce Fig. 2.8 with the DUPLICATE FRAME NUMBERS sorted into a random order:

Fig. 2.8 Duplicate Frame Numbers Sorted by Random Number

FRAME NO.	DUPLICATE FRAME NO.	RANDOM NO.
1	5	0.063981403
2	6	0.977468743
3	15	0.225170263
4	16	0.765734052
5	32	0.274680922
6	1	0.594468001
7	26	0.511966171
8	28	0.625577233
9	24	0.906310053
10	11	0.488640116
11	2	0.020129977
12	25	0.723003676
13	8	0.975227547
14	7	0.469582962
15	9	0.14889954
16	22	0.955629903
17	10	0.897398234
18	3	0.314860892
19	19	0.442019486
20	30	0.078566335
21	29	0.172474705
22	27	0.104689528
23	17	0.406630369
24	21	0.961398315
25	23	0.094222677
26	13	0.323429051
27	20	0.470615753
28	12	0.978014724
29	4	0.618082813
30	14	0.727776384
31	18	0.919475329
32	31	0.324497007

Important note: Because Excel randomly assigns these random numbers, your Excel commands will produce a different sequence of random numbers from everyone else who reads this book!

Because your objective at the beginning of this chapter was to select randomly 5 of the 32 areas of the salt marsh, you now can do that by selecting the *first five ID numbers* in DUPLICATE FRAME NO. column after the sort.

Although your first five random numbers will be different from those we have selected in the random sort that we did in this chapter, we would select these five IDs of areas using Fig. 2.9.

Fig. 2.9 First Five Areas Selected Randomly

FRAME NO.	DUPLICATE FRAME NO.	RANDOM NO.
1	5	0.063981403
2	6	0.977468743
3	15	0.225170263
4	16	0.765734052
5	32	0.274680922
6	1	0.594468001
7	26	0.511966171
8	28	0.625577233
9	24	0.906310053
10	11	0.488640116
11	2	0.020129977
12	25	0.723003676
13	8	0.975227547
14	7	0.469582962
15	9	0.14889954
16	22	0.955629903
17	10	0.897398234
18	3	0.314860892
19	19	0.442019486
20	30	0.078566335
21	29	0.172474705
22	27	0.104689528
23	17	0.406630369
24	21	0.961398315
25	23	0.094222677
26	13	0.323429051
27	20	0.470615753
28	12	0.978014724
29	4	0.618082813
30	14	0.727776384
31	18	0.919475329
32	31	0.324497007

5, 6, 15, 16, 32
Save this file as: Random33

Remember, your five ID numbers selected after your random sort will be different from the five ID numbers in Fig. 2.9 because Excel assigns a different random number *each time the* = *RAND() command is given.*

Before we leave this chapter, you need to learn how to print a file so that all of the information on that file fits onto a single page without "dribbling over" onto a second or third page.

2.4 Printing an Excel File So That All of the Information Fits Onto One Page

Objective: To print a file so that all of the information fits onto one page

Note that the three practice problems at the end of this chapter require you to sort random numbers when the files contain 63 resistors, 114 steel samples, and 75 toxic waste sites, respectively. These files will be "too big" to fit onto one page when you print them unless you format these files so that they fit onto a single page when you print them.

Let's create a situation where the file does not fit onto one printed page unless you format it first to do that.

Go back to the file you just created, Random 33, and enter the name: *Jennifer* into cell: A50.

If you printed this file now, the name, *Jennifer*, would be printed onto a second page because it "dribbles over" outside of the page rage for this file in its current format.

So, you would need to change the page format so that all of the information, including the name, Jennifer, fits onto just one page when you print this file by using the following steps:

Page Layout (top left of the computer screen)
(Notice the "Scale to Fit" section in the center of your screen; see Fig. 2.10)

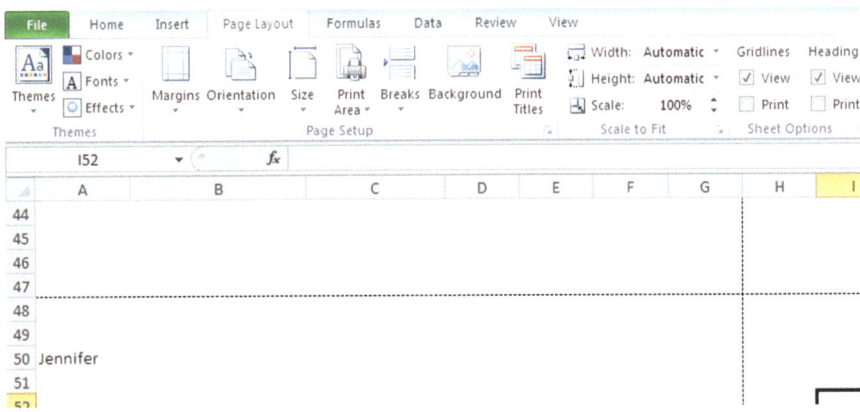

Fig. 2.10 Dialogue Box for Page Layout / Scale to Fit Commands

Hit the down arrow to the right of 100 % *once* to reduce the size of the page to 95 %

Now, note that the name, Jennifer, is still on a second page on your screen because her name is below the horizontal dotted line on your screen in Fig. 2.11 (the dotted lines tell you outline dimensions of the file if you printed it now).

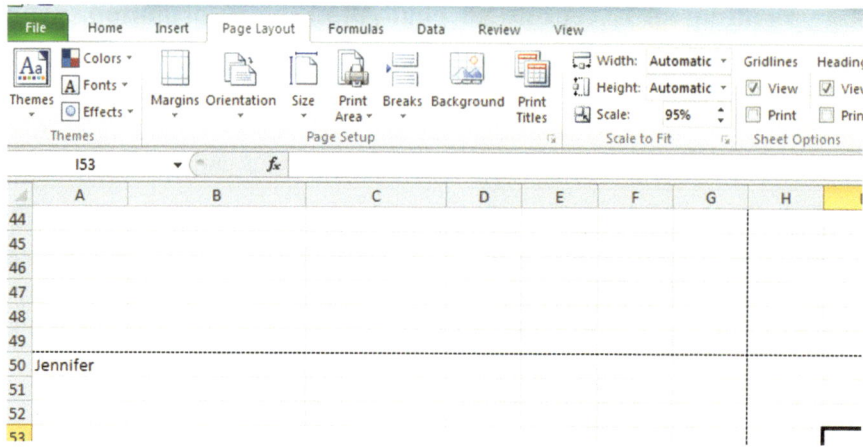

Fig. 2.11 Example of Scale Reduced to 95 % with "Jennifer" to be Printed on a Second Page

So, you need to repeat the "scale change steps" by hitting the down arrow on the right once more to reduce the size of the worksheet to 90 % of its normal size.

Notice that the "dotted lines" on your computer screen in Fig. 2.12 are now below Jennifer's name to indicate that all of the information, including her name, is now formatted to fit onto just one page when you print this file.

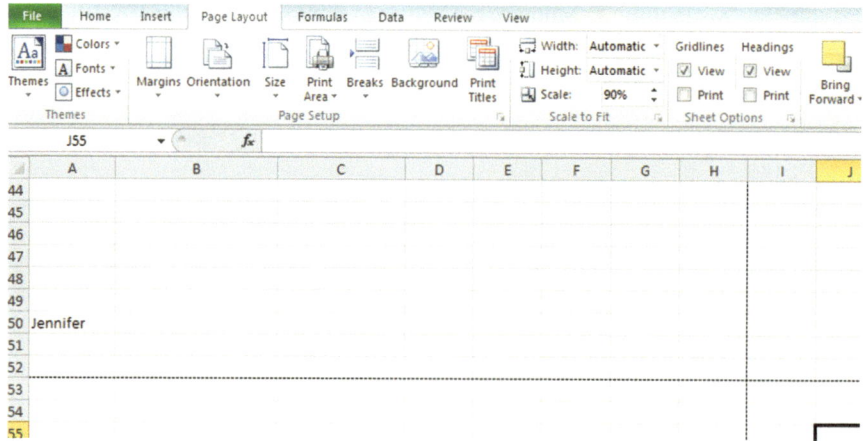

Fig. 2.12 Example of Scale Reduced to 90 % with "Jennifer" to be printed on the first page (note the dotted line below Jennifer on your screen)

Save the file as: Random34

Print the file. Does it all fit onto one page? It should (see Fig. 2.13).

Fig. 2.13 Final Spreadsheet
of 90 % Scale to Fit

FRAME NO.	DUPLICATE FRAME NO.	RANDOM NO.
1	5	0.747176905
2	6	0.038774393
3	15	0.091368861
4	16	0.63147137
5	32	0.190734495
6	1	0.411943765
7	26	0.138033007
8	28	0.927874602
9	24	0.058336576
10	11	0.043243606
11	2	0.729011126
12	25	0.204119693
13	8	0.456656709
14	7	0.232589896
15	9	0.09096704
16	22	0.935399501
17	10	0.201267198
18	3	0.52638312
19	19	0.53734605
20	30	0.969840616
21	29	0.475657455
22	27	0.558049277
23	17	0.488444809
24	21	0.717097206
25	23	0.86192944
26	13	0.875595013
27	20	0.536748908
28	12	0.331784725
29	4	0.642847666
30	14	0.575767804
31	18	0.939789757
32	31	0.776050794

Jennifer

2.5   End-of-Chapter Practice Problems

1. Suppose that you work for an electronics company and that you needed to test electrical resistors for quality purposes. You have 63 resistors of one particular type. You need to randomly test 15 of these 63 resistors.

   (a) Set up a spreadsheet of frame numbers for these resistors with the heading: FRAME NUMBERS using the Home/Fill commands.
   (b) Then, create a separate column to the right of these frame numbers which duplicates these frame numbers with the title: Duplicate frame numbers
   (c) Then, create a separate column to the right of these duplicate frame numbers and use the =RAND() function to assign random numbers to all of the frame numbers in the duplicate frame numbers column, and change this column format so that 3 decimal places appear for each random number
   (d) Sort the duplicate frame numbers and random numbers into a random order
   (e) Print the result so that the spreadsheet fits onto one page
   (f) Circle on your printout the I.D. number of the first 15 resistors that you would use in your research study
   (g) Save the file as: RAND9

   > *Important note: Note that everyone who does this problem will generate a different random order of resistor ID numbers since Excel assigns a different random number each time the RAND() command is used. For this reason, the answer to this problem given in this Excel Guide will have a completely different sequence of random numbers from the random sequence that you generate. This is normal and what is to be expected.*

2. Suppose that you have been hired as a consultant to test building materials for engineers designing suspension bridges. The engineers of this company are interested in using a new type of steel in future bridge construction. You have been given 114 samples of this type of steel and you have been asked to test a random sample of 10 of these samples for tensile strength in terms of their material consistency.

   (a) Set up a spreadsheet of frame numbers for these steel samples with the heading: FRAME NO.
   (b) Then, create a separate column to the right of these frame numbers which duplicates these frame numbers with the title: Duplicate frame no.
   (c) Then, create a separate column to the right of these duplicate frame numbers entitled "Random number" and use the=RAND() function to assign random numbers to all of the frame numbers in the duplicate frame numbers column. Then, change this column format so that 3 decimal places appear for each random number
   (d) Sort the duplicate frame numbers and random numbers into a random order

(e) Print the result so that the spreadsheet fits onto one page
(f) Circle on your printout the I.D. number of the first 10 steel samples that would be used in this research study.
(g) Save the file as: RANDOM6

3. Suppose that a chemical field researcher wants to take a random sample of 20 of 75 toxic waste sites that have been mapped surrounding a commercial house paint plant that has been closed and abandoned. The researcher wants to test the amount of lead in the soil around this plant as part of a field research study.

(a) Set up a spreadsheet of frame numbers for these sites with the heading: FRAME NUMBERS.
(b) Then, create a separate column to the right of these frame numbers which duplicates these frame numbers with the title: Duplicate frame numbers
(c) Then, create a separate column to the right of these duplicate frame numbers entitled "Random number" and use the =RAND() function to assign random numbers to all of the frame numbers in the duplicate frame numbers column. Then, change this column format so that 3 decimal places appear for each random number
(d) Sort the duplicate frame numbers and random numbers into a random order
(e) Print the result so that the spreadsheet fits onto one page
(f) Circle on your printout the I.D. number of the first 20 sites that the field chemist should select for her study.
(g) Save the file as: RAND5

Chapter 3
Confidence Interval About the Mean Using the TINV Function and Hypothesis Testing

This chapter focuses on two ideas: (1) finding the 95 % confidence interval about the mean, and (2) hypothesis testing.

Let's talk about the confidence interval first.

3.1 Confidence Interval About the Mean

In statistics, we are always interested in *estimating the population mean*. How do we do that?

3.1.1 How to Estimate the Population Mean

Objective: To estimate the population mean, μ

Remember that the population mean is the average of all of the people in the target population. For example, if we were interested in how well adults ages 25–44 liked a new flavor of Ben & Jerry's ice cream, we could never ask this question of all of the people in the U.S. who were in that age group. Such a research study would take way too much time to complete and the cost of doing that study would be prohibitive.

So, instead of testing *everyone* in the population, we take a sample of people in the population and use the results of this sample to estimate the mean of the entire population. This saves both time and money. When we use the results of a sample to estimate the population mean, this is called "*inferential statistics*" because we are inferring the population mean from the sample mean.

T.J. Quirk et al., *Excel 2010 for Physical Sciences Statistics: A Guide to Solving Practical Problems*, DOI 10.1007/978-3-319-00630-7_3,
© Springer International Publishing Switzerland 2013

When we study a sample of people in science research, we know the size of our sample (n), the mean of our sample ($\overline{X}$), and the standard deviation of our sample (STDEV). We use these figures to estimate the population mean with a test called the "confidence interval about the mean."

3.1.2   Estimating the Lower Limit and the Upper Limit of the 95 Percent Confidence Interval About the Mean

The theoretical background of this test is beyond the scope of this book, and you can learn more about this test from studying any good statistics textbook (e.g. McKillup and Dyar 2010 or Ledolter and Hogg 2010), but the basic ideas are as follows.

We assume that the population mean is somewhere in an interval which has a "lower limit" and an "upper limit" to it. We also assume in this book that we want to be "95 % confident" that the population mean is inside this interval somewhere. So, we intend to make the following type of statement:

"We are 95 % confident that the population mean in miles per gallon (mpg) for the Chevy Impala automobile is between 26.92 miles per gallon and 29.42 miles per gallon."

If we want to create a billboard emphasing the perceived lower environmental impact of the Chevy Impala by claiming that this car gets 28 miles per gallon (mpg), we can do this because 28 is *inside the 95 % confidence interval* in our research study in the above example. We do not know exactly what the population mean is, only that it is somewhere between 26.92 mpg and 29.42 mpg, and 28 is inside this interval.

But we are only 95 % confident that the population mean is inside this interval, and 5 % of the time we will be wrong in assuming that the population mean is 28 mpg.

But, for our purposes in science research, we are happy to be 95 % confident that our assumption is accurate. We should also point out that 95 % is an arbitrary level of confidence for our results. We could choose to be 80 % confident, or 90 % confident, or even 99 % confident in our results if we wanted to do that. But, in this book, *we will always assume that we want to be 95 % confident of our results.* That way, you will not have to guess on how confident you want to be in any of the problems in this book. We will always want to be 95 % confident of our results in this book.

So how do we find the 95 % confidence interval about the mean for our data? In words, we will find this interval this way:

"Take the sample mean ($\overline{X}$), *and add to it* 1.96 times the standard error of the mean (s.e.) to get the upper limit of the confidence interval. Then, take the sample mean, *and subtract from it* 1.96 times the standard error of the mean to get the lower limit of the confidence interval."

You will remember (See Section 1.3) that the standard error of the mean (s.e.) is found by dividing the standard deviation of our sample (STDEV) by the square root of our sample size, n.

In mathematical terms, the formula for the 95 % confidence interval about the mean is:

$$\overline{X} \pm 1.96\,\text{s.e.} \tag{3.1}$$

Note that the "$\pm$ *sign*" stands for "plus or minus," and this means that you first add 1.96 times the s.e. to the mean to get the upper limit of the confidence interval, and then subtract 1.96 times the s.e. from the mean to get the lower limit of the confidence interval. Also, the symbol 1.96 s.e. means that you multiply 1.96 times the standard error of the mean to get this part of the formula for the confidence interval.

Note: We will explain shortly where the number 1.96 came from.

Let's try a simple example to illustrate this formula.

3.1.3   Estimating the Confidence Interval for the Chevy Impala in Miles Per Gallon

Let's suppose that you have been asked to be a part of a larger study looking at the carbon footprint of Chevy Impala drivers. You are interested in the average miles per gallon (mpg) of a Chevy Impala. You asked owners of the Chevy Impala to keep track of their mileage and the number of gallons used for two tanks of gas. Let's suppose that 49 owners did this, and that they average 27.83 miles per gallon (mpg) with a standard deviation of 3.01 mpg. The standard error (s.e.) would be 3.01 divided by the square root of 49 (i.e., 7) which gives a s.e. equal to 0.43.

The 95 % confidence interval for these data would be:

$$27.83 \pm 1.96\,(0.43)$$

The *upper limit of this confidence interval* uses the plus sign of the $\pm$ sign in the formula. Therefore, the upper limit would be:

$$27.83 + 1.96\,(0.43) = 27.83 + 0.84 = 28.67 \text{ mpg}$$

Similarly, *the lower limit of this confidence interval* uses the minus sign of the $\pm$ sign in the formula. Therefore, the lower limit would be:

$$27.83 - 1.96\,(0.43) = 27.83 - 0.84 = 26.99 \text{ mpg}$$

The result of our part of the ongoing research study would, therefore, be the following:

"We are 95 % confident that the population mean for the Chevy Impala is somewhere between 26.99 mpg and 28.67 mpg."

Based upon the 28 mpg of the Chevy Impala being inside the confidence interval, we could create a billboard emphasizing the 28 miles per gallon and highlight a perceived lower environmental impact. Our data supports this claim because the 28 mpg is inside of this 95 % confidence interval for the population mean.

You are probably asking yourself: "Where did that 1.96 in the formula come from?"

3.1.4 Where Did the Number "1.96" Come From?

A detailed mathematical answer to that question is beyond the scope of this book, but here is the basic idea.

We make an assumption that the data in the population are "normally distributed" in the sense that the population data would take the shape of a "normal curve" if we could test all of the people or properties in the population. The normal curve looks like the outline of the Liberty Bell that sits in front of Independence Hall in Philadelphia, Pennsylvania. The normal curve is "symmetric" in the sense that if we cut it down the middle, and folded it over to one side, the half that we folded over would fit perfectly onto the half on the other side.

A discussion of integral calculus is beyond the scope of this book, but essentially we want to find the lower limit and the upper limit of the population data in the normal curve so that 95 % of the area under this curve is between these two limits. *If we have more than 40 people in our research study*, the value of these limits is plus or minus 1.96 times the standard error of the mean (s.e.) of our sample. The number 1.96 times the s.e. of our sample gives us the upper limit and the lower limit of our confidence interval. If you want to learn more about this idea, you can consult a good statistics book (e.g. McKillup and Dyar 2010).

The number 1.96 would change if we wanted to be confident of our results at a different level from 95 % as long as we have more than 40 people in our research study.

For example:

1. If we wanted to be 80 % confident of our results, this number would be 1.282.
2. If we wanted to be 90 % confident of our results, this number would be 1.645.
3. If we wanted to be 99 % confident of our results, this number would be 2.576.

But since we always want to be 95 % confident of our results in this book, we will always use 1.96 in this book whenever we have more than 40 people in our research study.

By now, you are probably asking yourself: "Is this number in the confidence interval about the mean always 1.96 ?" The answer is: "No!", and we will explain why this is true now.

3.1.5 Finding the Value for t in the Confidence Interval Formula

Objective: To find the value for t in the confidence interval formula

The correct formula for the confidence interval about the mean for different sample sizes is the following:

$$\overline{X} \ \pm \ t \ \text{s.e.} \tag{3.2}$$

To use this formula, you find the sample mean, $\overline{X}$, *and add to it the value of t times the s.e. to get the upper limit* of this 95 % confidence interval. Also, you take the sample mean, $\overline{X}$, and *subtract from it the value of t times the s.e. to get the lower limit* of this 95 % confidence interval. And, you find the value of t in the table given in Appendix E of this book in the following way:

Objective: To find the value of t in the t-table in Appendix E

Before we get into an explanation of what is meant by "the value of t," let's give you practice in finding the value of t by using the t-table in Appendix E.

Keep your finger on Appendix E as we explain how you need to "read" that table.

Since the test in this chapter is called the "confidence interval about the mean test," you will use the first column on the left in Appendix E to find the critical value of t for your research study (note that this column is headed: "sample size n").

To find the value of t, you go down this first column until you find the sample size in your research study, and then you go to the right and read the value of t for that sample size in the "critical t column" of the table (note that this column is the column that you would use for the 95 % confidence interval about the mean).

For example, if you have 14 people in your research study, the value of t is 2.160. If you have 26 people in your research study, the value of t is 2.060. If you have more than 40 people in your research study, the value of t is always 1.96.

Note that the "critical t column" in Appendix E represents the value of t that you need to use to obtain to be 95 % confident of your results as "significant" results.

Throughout this book, we are assuming that you want to be 95 % confident in the results of your statistical tests. Therefore, the value for t in the t-table in Appendix E tells you which value you should use for t when you use the formula for the 95 % confidence interval about the mean.

Now that you know how to find the value of t in the formula for the confidence interval about the mean, let's explore how you find this confidence interval using Excel.

3.1.6 Using Excel's TINV Function to Find the Confidence Interval About the Mean

---
Objective: To use the TINV function in Excel to find the confidence interval
          about the mean

When you use Excel, the formulas for finding the confidence interval are:

$Lower\ limit: = \overline{X} - TINV(1 - 0.95, n - 1)^{*}s.e.$ (no spaces between these symbols)

$$(3.3)$$

$Upper\ limit: = \overline{X} + TINV(1 - 0.95, n - 1)^{*}s.e.$ (no spaces between these symbols)

$$(3.4)$$

Note that the "*symbol" in this formula tells Excel to use the multiplication step in the formula, and it stands for "times" in the way we talk about multiplication.

You will recall from Chapter 1 that n stands for the sample size, and so $n - 1$ stands for the sample size minus one.

You will also recall from Chapter 1 that the standard error of the mean, s.e., equals the STDEV divided by the square root of the sample size, n (See Section 1.3).

Let's try a sample problem using Excel to find the 95 % confidence interval about the mean for a problem.

Let's suppose that General Motors wanted to claim that its Chevy Impala achieves 28 miles per gallon (mpg). Let's call 28 mpg the "reference value" for this car.

Suppose that you work for Ford Motor Co. and that you want to check this claim to see is it holds up based on some research evidence. You decide to collect some data and to use a two-side 95 % confidence interval about the mean to test your results:

3.1.7 Using Excel to find the 95 Percent Confidence Interval for a Car's mpg Claim

---
Objective: To analyze the data using a two-side 95 % confidence interval about
          the mean

You select a sample of new car owners for this car and they agree to keep track of their mileage for two tanks of gas and to record the average miles per gallon they achieve on these two tanks of gas. Your research study produces the hypothetical results given in Fig. 3.1:

Chevy Impala

Miles per gallon
30.9
24.5
31.2
28.7
35.1
29.0
28.8
23.1
31.0
30.2
28.4
29.3
24.2
27.0
26.7
31.0
23.5
29.4
26.3
27.5
28.2
28.4
29.1
21.9
30.9

Fig. 3.1 Worksheet Data for Chevy Impala (Practical Example)

Create a spreadsheet with these data and use Excel to find the sample size (n), the mean, the standard deviation (STDEV), and the standard error of the mean (s.e.) for these data using the following cell references.

A3:   Chevy Impala
A5:   Miles per gallon
A6   30.9

Enter the other mpg data in cells A7: A30

Now, highlight cells A6:A30 and format these numbers in number format (one decimal place). Center these numbers in Column A. Then, widen columns A and

B by making both of them twice as wide as the original width of column
A. Then, widen column C so that it is three times as wide as the original width of
column A so that your table looks more professional.

C7:   n
C10:   Mean
C13:   STDEV
C16:   s.e.
C19:   95 % confidence interval
D21:   Lower limit:
D23:   Upper limit: (see Fig. 3.2)

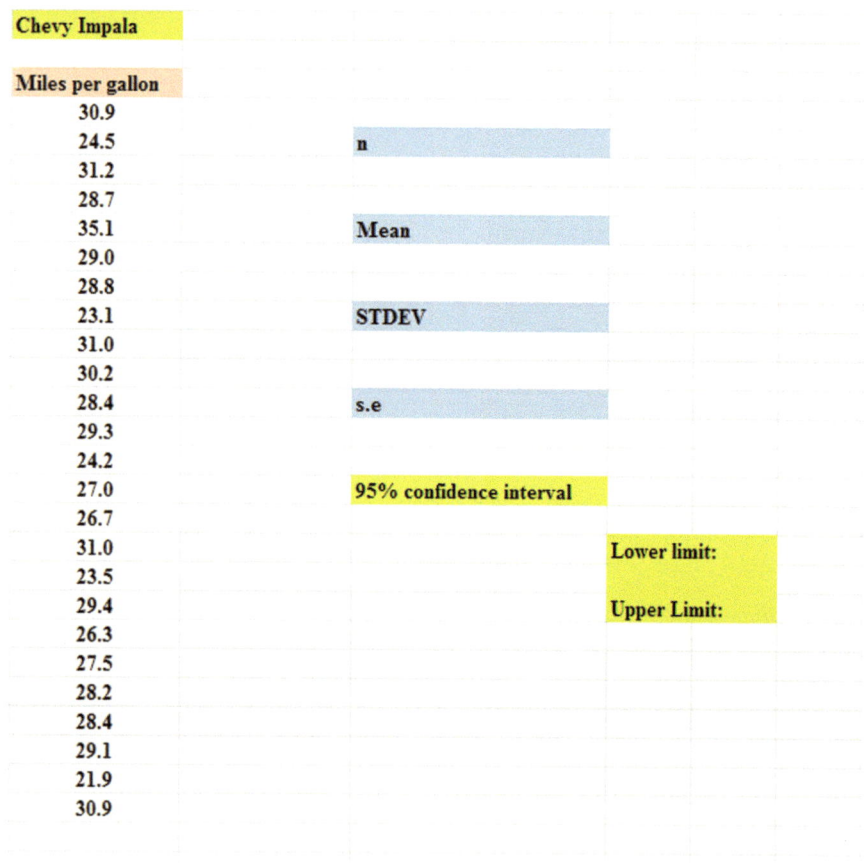

Fig. 3.2 Example of Chevy Impala Format for the Confidence Interval About the Mean Labels

B26:   Draw a picture below this confidence interval
B28:   26.92
B29:   lower (right-align this word)
B30:   limit (right-align this word)
C28:   '------28 -------28.17 -------- (note that you need to begin cell C28 with a
       single quotation mark (') to tell Excel that this is a *label*, and not a number)
D28:   '------------- (note the single quotation mark)
E28:   '29.42 (note the single quotation mark)
C29:   ref.   Mean
C30:   value
E29:   upper
E30:   limit
B33:   Conclusion:

Now, align the labels underneath the picture of the confidence interval so that they look like Figure 3.3.

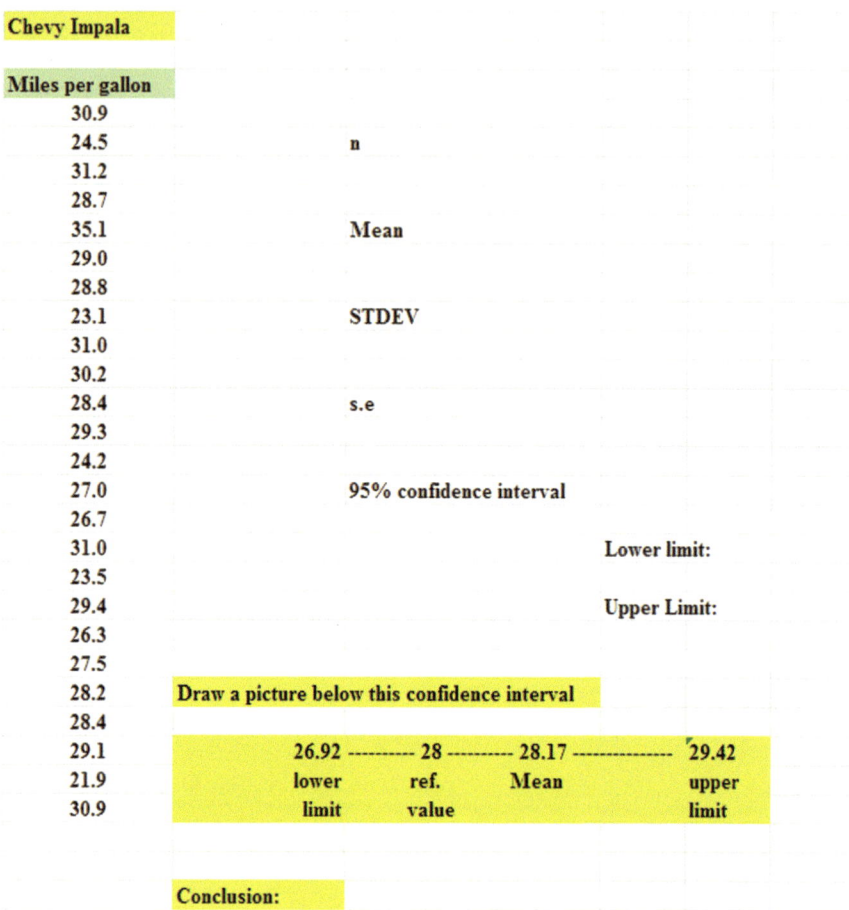

Fig. 3.3 Example of Drawing a Picture of a Confidence Interval About the Mean Result

Next, name the range of data from A6:A30 as: miles

D7:   Use Excel to find the sample size
D10:   Use Excel to find the mean
D13:   Use Excel to find the STDEV
D16:   Use Excel to find the s.e.

Now, you need to find the lower limit and the upper limit of the 95 % confidence interval for this study.

We will use Excel's TINV function to do this. We will assume that you want to be 95 % confident of your results.

F21:   =D10−TINV(1−.95,24)*D16

Note that this TINV formula uses 24 since 24 is one less than the sample size of 25 (i.e., 24 is n−1). Note that D10 is the mean, while D16 is the standard error of the mean. The above formula gives the *lower limit of the confidence interval, 26.92.*

F23:   =D10+TINV(1−.95,24)*D16

The above formula gives the *upper limit of the confidence interval, 29.42.*

Now, use number format (two decimal places) in your Excel spreadsheet for the mean, standard deviation, standard error of the mean, and for both the lower limit and the upper limit of your confidence interval. If you printed this spreadsheet now, the lower limit of the confidence interval (26.92) and the upper limit of confidence interval (29.42) would "dribble over" onto a second printed page because the information on the spreadsheet is too large to fit onto one page in its present format.

So, you need to use Excel's "Scale to Fit" commands that we discussed in Chapter 2 (see Sect. 2.4) to reduce the size of the spreadsheet to 95 % of its current size using the Page Layout/Scale to Fit function. Do that now, and notice that the dotted line to the right of 26.92 and 29.42 indicates that these numbers would now fit onto one page when the spreadsheet is printed out (see Fig. 3.4)

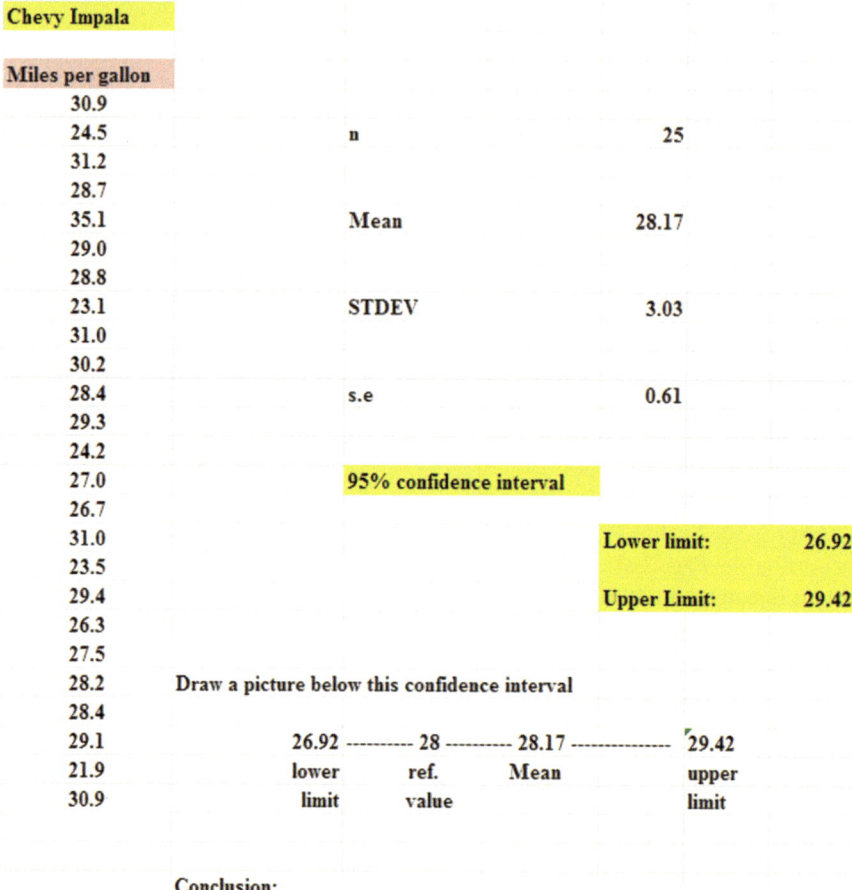

Fig. 3.4 Result of Using the TINV Function to Find the Confidence Interval About the Mean

Note that you have drawn a picture of the 95 % confidence interval beneath cell B26, including the lower limit, the upper limit, the mean, and the reference value of 28 mpg given in the claim that the company wants to make about the car's miles per gallon performance.

Now, let's write the conclusion to your research study on your spreadsheet:

C33:   Since the reference value of 28 is inside
C34:   the confidence interval, we accept that
C35:   the Chevy Impala does get 28 mpg.

Important note: You are probably wondering why we wrote the conclusion on three separatelines of the spreadsheet instead of writing it on one long line. This is because if you wrote it on one line, two things would happen that you would not like: (1) If you printed the conclusion by reducing the size of the layout of the page so that the entire spreadsheet would fit onto one page, the print font size for the entire spreadsheet would be so small that you could not read it without a magnifying glass, and (2) If you printed the spreadsheet without reducing the page size layout, it would "dribble over" part of the conclusion to a separate page all by itself, and your spreadsheet would not look professional.

Your research study accepted the claim that the Chevy Impala did get 28 miles per gallon. The average miles per gallon in your study was 28.17. (See Fig. 3.5)

Save your resulting spreadsheet as: **CHEVY7**

Chevy Impala

Miles per gallon

30.9				
24.5		n		25
31.2				
28.7				
35.1		Mean		28.17
29.0				
28.8				
23.1		STDEV		3.03
31.0				
30.2				
28.4		s.e		0.61
29.3				
24.2				
27.0		95% confidence interval		
26.7				
31.0			Lower limit:	26.92
23.5				
29.4			Upper Limit:	29.42
26.3				
27.5				
28.2		Draw a picture below this confidence interval		
28.4				
29.1		26.92 ---------- 28 ---------- 28.17 -------------- 29.42		
21.9		lower ref. Mean upper		
30.9		limit value limit		

Conclusion: **Since the reference value of 28 is inside the confidence interval, we accept that the Chevy Impala does get 28 mpg.**

Fig. 3.5 Final Spreadsheet for the Chevy Impala Confidence Interval About the Mean

3.2   Hypothesis testing

One of the important activities of research scientists is that they attempt to "check" their assumptions about the world by testing these assumptions in the form of hypotheses.

A typical hypothesis is in the form: *"If x, then y."*

Some examples would be:

1. "If we use this new method fertilizing the soil, the corn yield of the plot will increase by 3 percent."
2. "There will be no difference in the concentration of lead in water that has circulated in soils mixed with apatite (a type of mineral found in rocks that is used to make fertilizer) compared to soils without apatite."
3. "If we change the format for teaching Introductory Chemistry to our undergraduates, then their final exam scores will increase by 8 percent."

A hypothesis, then, to a research scientist is a "guess" about what we think is true in the real world. We can test these guesses using statistical formulas to see if our predictions come true in the real world.

So, in order to perform these statistical tests, we must first state our hypotheses so that we can test our results against our hypotheses to see if our hypotheses match reality.

So, how do we generate hypotheses in science research?

3.2.1   Hypotheses Always Refer to the Population of Physical Properties that You Are Studying

The first step is to understand that our hypotheses always refer to the *population* of physical properties in a study.

For example, suppose we are interested in studying the brightness of various types of light bulbs used in a certain type of vehicle headlight. We would select various types of light bulbs used in the vehicle headlight and measure the brightness (in lumens) of each bulb type. These brightness measurements would be used as our sample. This sample would be used in generalizing our findings for all of the light bulbs used in this vehicle.

All of the light bulbs used in this type of vehicle would be the *population* that we are interested in studying, while the specific light bulbs in our study are called the *sample* from this population.

Since our sample sizes typically contain only a portion of the light bulbs, we are interested in the results of our sample *only insofar as the results of our sample can be "generalized" to the population in which we are really interested.*

That is why our hypotheses always refer to the population, and never to the sample of physical properties in our study.

You will recall from Chapter 1 that we used the symbol: $\overline{X}$ to refer to the mean of the sample we use in our research study (See Section 1.1).

We will use the symbol: μ (the Greek letter "mu") to refer to the *population mean.*

In testing our hypotheses, we are trying to decide which one of two competing hypotheses *about the population mean* we should accept given our data set.

3.2.2 The Null Hypothesis and the Research (Alternative) Hypothesis

These two hypotheses are called the *null hypothesis* and the *research hypothesis.*

Statistics textbooks typically refer to the *null hypothesis* with the notation: H_0.

The *research hypothesis* is typically referred to with the notation: H_1, and it is sometimes called the *alternative hypothesis.*

Let's explain first what is meant by the null hypothesis and the research hypothesis:

(1) *The null hypothesis is what we accept as true unless we have compelling evidence that it is not true.*
(2) *The research hypothesis is what we accept as true whenever we reject the null hypothesis as true.*

This is similar to our legal system in America where we assume that a supposed criminal is innocent until he or she is proven guilty in the eyes of a jury. Our null hypothesis is that this defendant is innocent, while the research hypothesis is that he or she is guilty.

In the great state of Missouri, every license plate has the state slogan: "Show me." This means that people in Missouri think of themselves as not gullible enough to accept everything that someone says as true unless that person's actions indicate the truth of his or her claim. In other words, people in Missouri believe strongly that a person's actions speak much louder than that person's words.

Since both the null hypothesis and the research hypothesis cannot both be true, the task of hypothesis testing using statistical formulas is to decide which one you will accept as true, and which one you will reject as true.

Sometimes in science research a series of rating scales is used to measure people's attitudes toward a company, toward one of its products, or toward their intention-to-buy that company's products. These rating scales are typically 5-point, 7-point, or 10-point scales, although other scale values are often used as well.

3.2.2.1 Determining the Null Hypothesis and the Research Hypothesis When Rating Scales are Used

The following examples are another way to test the null hypothesis and research hypothesis using rating scales. Although rating scales are seldom used in the physical sciences, the following examples are good examples of how you would test these hypotheses if you encountered rating scales in your work.

Here is a typical example of a 7-point scale in science education for parents of 10th grade pupils at the end of a school year (see Fig. 3.6):

Overall, how satisfied are you with the quality of the academic program offered by your son's or daughter's school?

1	2	3	4	5	6	7
very dissatisfied						very satisfied

Null hypothesis: $\mu \ =$ _____

Research hypothesis: $\mu \ \neq$ _____

Fig. 3.6 Example of a Rating Scale Item for Parents of 10th Graders (Practical Example)

So, how do we decide what to use as the null hypothesis and the research hypothesis whenever rating scales are used?

Objective: To decide on the null hypothesis and the research hypothesis whenever rating scales are used.

In order to make this determination, we will use a simple rule:

Rule: Whenever rating scales are used, we will use the "middle" of the scale as the null hypothesis and the research hypothesis.

In the above example, since 4 is the number in the middle of the scale (i.e., three numbers are below it, and three numbers are above it), our hypotheses become:

Null hypothesis: $\mu = 4$
Research hypothesis: $\mu \neq 4$

In the above rating scale example, if the result of our statistical test for this one attitude scale item indicates that our population mean is "close to 4," we say that we accept the null hypothesis that the parents of 10th grade pupils were neither satisfied nor dissatisfied with the quality of the science program offered by their son's or daughter's school.

In the above example, *if the result of our statistical test indicates that the population mean is significantly different from 4*, we reject the null hypothesis and accept the research hypothesis *by stating either that*:

"Parents of 10th grade pupils were significantly satisfied with the quality of the science program offered by their son's or daughter's school" (this is true whenever our sample mean is significantly greater than our expected population mean of 4).

or

"Parents of 10th grade pupils were significantly dissatisfied with the quality of the science program offered by their son's or daughter's school" (this is accepted as true whenever our sample mean is significantly less than our expected population mean of 4).

Both of these conclusions cannot be true. We accept one of the hypotheses as "true" based on the data set in our research study, and the other one as "not true" based on our data set.

The job of the research scientist, then, is to decide which of these two hypotheses, the null hypothesis or the research hypothesis, he or she will accept as true given the data set in the research study.

Let's try some examples of rating scales so that you can practice figuring out what the null hypothesis and the research hypothesis are for each rating scale.

In the spaces in Fig. 3.7, write in the null hypothesis and the research hypothesis for the rating scales:

How did you do?

1. State University is an excellent university.

1	2	3	4	5
Strongly Disagree	Disagree	Undecided	Agree	Strongly Agree

Null hypothesis: μ = _____

Research hypothesis: μ ≠ _____

2. How would you rate the quality of teaching in the Physics Department at State University?

poor   1   2   3   4   5   6   7   excellent

Null hypothesis: μ = _____

Research hypothesis: μ ≠ _____

3. How would you rate the quality of the faculty in the Physics Department at State University?

1	2	3	4	5	6	7	8	9	10
very poor									very good

Null hypothesis: μ = _____

Research hypothesis: μ ≠ _____

Fig. 3.7 Examples of Rating Scales for Determining the Null Hypothesis and the Research Hypothesis

Here are the answers to these three questions:

1. The null hypothesis is $\mu = 3$, and the research hypothesis is $\mu \neq 3$ on this 5-point scale (i.e. the "middle" of the scale is 3).
2. The null hypothesis is $\mu = 4$, and the research hypothesis is $\mu \neq 4$ on this 7-point scale (i.e., the "middle" of the scale is 4).
3. The null hypothesis is $\mu = 5.5$, and the research hypothesis is $\mu \neq 5.5$ on this 10-point scale (i.e., the "middle" of the scale is 5.5 since there are 5 numbers below 5.5 and 5 numbers above 5.5).

As another example, Webster University, whose main campus is in St. Louis, Missouri USA, uses a Course Feedback form for student evaluations at the end of its courses which have 12 rating items referring to the course's planning and organization and the communications from the instructor to the students. The ratings are summarized and the results given to instructors after the course is completed. Each of these items is rated on the following 4-point scale:

1 = Very Effective
2 = Effective
3 = Ineffective
4 = Very Ineffective

On this scale, the null hypothesis is: $\mu = 2.5$ and the research hypothesis is: $\mu \neq 2.5$, because there are two numbers below 2.5, and two numbers above 2.5 on these rating scales. (Note that the scalesare scored so that a low score, like a low score in golf, is a better score).

Now, let's discuss the 7 STEPS of hypothesis testing for using the confidence interval about the mean.

3.2.3   The 7 Steps for Hypothesis-testing Using the Confidence Interval About the Mean

> Objective: To learn the 7 steps of hypothesis-testing using the confidence interval about the mean

There are seven basic steps of hypothesis-testing for this statistical test.

3.2.3.1   STEP 1: State the null hypothesis and the research hypothesis

If you are using numerical scales in your survey, you need to remember that these hypotheses refer to the "middle" of the numerical scale. For example, if you are using 7-point scales with 1 = poor and 7 = excellent, these hypotheses would refer to the middle of these scales and would be:

Null hypothesis H_0: $\mu = 4$
Research hypothesis H_1: $\mu \neq 4$

3.2.3.2   STEP 2: Select the appropriate statistical test

In this chapter we are studying the confidence interval about the mean, and so we will select that test.

3.2.3.3   STEP 3: Calculate the formula for the statistical test

You will recall (see Section 3.15) that the formula for the confidence interval about the mean is:

$$\overline{X} \pm t \text{ s.e.} \qquad\qquad (3.2)$$

We discussed the procedure for computing this formula for the confidence interval about the mean using Excel earlier in this chapter, and the steps involved in using that formula are:

1. Use Excel's =COUNT function to find the sample size.
2. Use Excel's =AVERAGE function to find the sample mean, $\overline{X}$.
3. Use Excel's =STDEV function to find the standard deviation, STDEV.
4. Find the standard error of the mean (s.e.) by dividing the standard deviation (STDEV) by the square root of the sample size, n.
5. Use Excel's TINV function to find the lower limit of the confidence interval.
6. Use Excel's TINV function to find the upper limit of the confidence interval.

3.2.3.4   STEP 4: *Draw a picture of the confidence interval about the mean, including the mean, the lower limit of the interval, the upper limit of the interval, and the reference value given in the null hypothesis, H_0* (We will explain Step 4 later in the chapter)

3.2.3.5   STEP 5: Decide on a decision rule

(a) *If the reference value is inside the confidence interval, accept the null hypothesis, H_0*
(b) *If the reference value is outside the confidence interval, reject the null hypothesis, H_0, and accept the research hypothesis, H_1*

3.2.3.6 STEP 6: State the result of your statistical test.

There are two possible results when you use the confidence interval about the mean, and only one of them can be accepted as "true." So your result would be one of the following:

Either: Since the reference value is inside the confidence interval, *we accept the null hypothesis, H_0*

Or: Since the reference value is outside the confidence interval, *we reject the null hypothesis, H_0, and accept the research hypothesis, H_1*

3.2.3.7 STEP 7: State the conclusion of your statistical test in plain English!

In practice, this is more difficult than it sounds because you are trying to summarize the result of your statistical test in simple English that is both concise and accurate so that someone who has never had a statistics course (such as your boss, perhaps) can understand the conclusion of your test. This is a difficult task, and we will give you lots of practice doing this last and most important step throughout this book.

> Objective: To write the conclusion of the confidence interval about the mean test.

Let's set some basic rules for stating the conclusion of a hypothesis test.

Rule #1: *Whenever you reject H_0 and accept H_1, you must use the word "significantly" in the conclusion to alert the reader that this test found an important result.*

Rule #2: *Create an outline in words of the "key terms" you want to include in your conclusion so that you do not forget to include some of them.*

Rule #3: *Write the conclusion in plain English so that the reader can understand it even if that reader has never taken a statistics course.*

Let's practice these rules using the Chevy Impala Excel spreadsheet that you created earlier in this chapter, but first we need to state the hypotheses for that car.

If General Motors wants to claim that the Chevy Impala gets 28 miles per gallon on a billboard ad, the hypotheses would be:

H_0 : $\mu = 28$ mpg
H_1 : $\mu \neq 28$ mpg

You will remember that the reference value of 28 mpg was inside the 95 % confidence interval about the mean for your data, so we would accept H_0 for the Chevy Impala that the car does get 28 mpg.

Objective: To state the result when you accept H_0

Result: *Since the reference value of 28 mpg is inside the confidence interval, we accept the null hypothesis, H_0*

Let's try our three rules now:

Objective: To write the conclusion when you accept H_0

Rule #1:   *Since the reference value was inside the confidence interval, we cannot use the word "significantly" in the conclusion. This is a basic rule we are using in this chapter for every problem.*

Rule #2:   The key terms in the conclusion would be:

  – Chevy Impala
  – reference value of 28 mpg

Rule #3:   The Chevy Impala did get 28 mpg.

The process of writing the conclusion when you accept H_0 is relatively straightforward since you put into words what you said when you wrote the null hypothesis.

However, the process of stating the conclusion when you reject H_0 and accept H_1 is more difficult, so let's practice writing that type of conclusion with three practice case examples:

Objective: To write the result and conclusion when you reject H_0

CASE #1: Suppose that an ad in *The Wall Street Journal* claimed that the Honda Accord Sedan gets 34 miles per gallon. The hypotheses would be:

H_0 : $\mu = 34$ mpg
H_1 : $\mu \neq 34$ mpg

Suppose that your research yields the following confidence interval:

```
30_____31_____32_____34
lower              Mean              upper             Ref.
limit                                limit             Value
```

Result: *Since the reference value is outside the confidence interval, we reject the null hypothesis and accept the research hypothesis*

The three rules for stating the conclusion would be:

Rule #1:   We must include the word "significantly" since the reference value of 34 is outside the confidence interval.

Rule #2:   The key terms would be:

      – Honda Accord Sedan
      – significantly
      – either "more than" or "less than"
      – and probably closer to

Rule #3:   The Honda Accord Sedan got significantly less than 34 mpg, and it was probably closer to 31 mpg.

Note that this conclusion says that the mpg was less than 34 mpg because the sample mean was only 31 mpg. Note, also, that when you find a significant result by rejecting the null hypothesis, *it is not sufficient to say only: "significantly less than 34 mpg,"* because that does not tell the reader "how much less than 34 mpg" the sample mean was from 34 mpg. To make the conclusion clear, you need to add: "probably closer to 31 mpg" since the sample mean was only 31 mpg.

CASE #2:   The density of a substance does not change significantly within the sample. In almost all cases, the density of a substance will be the same no matter how much or how little of the substance you have. This is particularly useful in identifying minerals. Density is calculated by taking the mass (g) of an object and dividing it by its volume (cm^3). For example, silver has a density of 10.49 g/cm^3. If you are given a substance and told that it is "pure silver," and the density of that substance is not 10.49 g/cm^3, then the substance you have is not pure silver, no matter what it looks like. The density of the substance could be more or less depending on what is mixed in with it. Science doesn't lie when it comes to density.

Let's suppose that you have been asked by a company to determine if a substance they purchased as being pure silver was actually pure silver. Suppose, further, that you have obtained 50 random samples of the material this company purchased and that you have calculated the density of each of these samples. You want to practice your data interpretation skills on the hypothetical data which produces the confidence interval below:

The hypotheses for this test would be:

H_0 : $\mu = 10.49$ g/cm^3
H_1 : $\mu \neq 10.49$ g/cm^3

Essentially, the null hypothesis equal to 10.49 g/cm^3 states that if the obtained mean score for this sample is not significantly different from 10.49 g/cm^3, then the substance purchased by the company was really pure silver.

Suppose that your analysis produced the following confidence interval for this test:

10.41_____10.43_____10.45_____10.49_____
lower   Mean   upper   Ref.
limit   limit   Value

Result: Since the reference value is outside the confidence interval, we reject the null hypothesis and accept the research hypothesis.

Rule #1:   You must include the word "significantly" since the reference value is outside the confidence interval

Rule #2:   The key terms would be:

   – density
   – substance tested
   – significantly
   – less or greater (depending on your result)
   – either pure silver or not pure silver (since the result is significant)

Rule #3:   The observed density of the substance tested was significantly less than the known density of silver. Therefore, the tested substance was not pure silver.

Note that you need to use the word "less" since the sample mean of 10.43 g/cm^3 was less than the reference value of 10.49 g/cm^3.

CASE #3: Your work as a "quality control supervisor" in a machine shop that produces steel rods of various diameters and lengths for use in construction projects. Your machine shop is known for producing high quality rods that consistently meet specific standards set by customers. Recently, the machine that cuts the rods to length at the shop has been having problems with making accurate cuts. The cutting machine has supposedly been "fixed," but management wants to make sure that the machine is now accurate. You have been asked to use your Excel skills to determine if the cutting machine is now working properly. You have obtained a random set of test rods that were cut with the machine when the machine was set to cut them to a length of 5.5 centimeters (cm).

The hypotheses would be:

H_0 : $\mu = 5.5$ cm
H_1 : $\mu \neq 5.5$ cm

Suppose that your research produced the following confidence interval for this machine for your test:

5.5_____5.7_____5.8_____5.9_____
Ref.   lower   Mean   upper
Value   limit   limit

Result: *Since the reference value is outside the confidence interval, we reject the null hypothesis and accept the research hypothesis*

The three rules for stating the conclusion would be:

Rule #1:   You must include the word "significantly" since the reference value is outside the confidence interval

Rule #2:   The key terms would be:

      – sample of test rods
      – significantly
      – longer or shorter (depending on the result of your test)

Rule #3:   The sample of test rods were cut significantly longer than what the cutting machine was set for at 5.5 cm, and were probably closer to 5.8 cm.

Note two important things about this conclusion above: (1) people when speaking English do not normally say "significantly longer" but your use of statistics allows you to speak with authority in this case, and (2) the mean of 5.8 cm was greater (longer) than the reference value of 5.5 cm.

If you want a more detailed explanation of the confidence interval about the mean, see Townend (2002).

The three practice problems at the end of this chapter will give you additional practice in stating the conclusion of your result, and this book will include many more examples that will help you to write a clear and accurate conclusion to your research findings.

3.3   Alternative Ways to Summarize the Result of a Hypothesis Test

It is important for you to understand that in this book we are summarizing an hypothesis test in one of two ways: (1) We accept the null hypothesis, or (2) We reject the null hypothesis and accept the research hypothesis. We are consistent in the use of these words so that you can understand the concept underlying hypothesis testing.

However, there are many other ways to summarize the result of an hypothesis test, and all of them are correct theoretically, even though the terminology differs. If you are taking a course with a professor who wants you to summarize the results of a statistical test of hypotheses in language which is different from the language we are using in this book, do not panic! If you understand the concept of hypothesis testing as described in this book, you can then translate your understanding to use the terms that your professor wants you to use to reach the same conclusion to the hypothesis test.

Statisticians and professors of science statistics all have their own language that they like to use to summarize the results of an hypothesis test. There is no one set of words that these statisticians and professors will ever agree on, and so we have chosen the one that we believe to be easier to understand in terms of the concept of hypothesis testing.

To convince you that there are many ways to summarize the results of an hypothesis test, we present the following quotes from prominent statistics and research books to give you an idea of the different ways that are possible.

3.3.1 Different Ways to Accept the Null Hypothesis

The following quotes are typical of the language used in statistics and research books *when the null hypothesis is accepted*:

"The null hypothesis is not rejected." (Black 2010, p. 310)
"The null hypothesis cannot be rejected." (McDaniel and Gates 2010, p. 545)
"The null hypothesis ... claims that there is no difference between groups." (Salkind 2010, p. 193)
"The difference is not statistically significant." (McDaniel and Gates 2010, p. 545)
" ... the obtained value is not extreme enough for us to say that the difference between Groups 1 and 2 occurred by anything other than chance." (Salkind 2010, p. 225)
"If we do not reject the null hypothesis, we conclude that there is not enough statistical evidence to infer that the alternative (hypothesis) is true." (Keller 2009, p. 358)
"The research hypothesis is not supported." (Zikmund and Babin 2010, p. 552)

3.3.2 Different Ways to Reject the Null Hypothesis

The following quotes are typical of the quotes used in statistics and research books *when the null hypothesis is rejected*:

"The null hypothesis is rejected." (McDaniel and Gates 2010, p. 546)
"If we reject the null hypothesis, we conclude that there is enough statistical evidence to infer that the alternative hypothesis is true." (Keller 2009, p. 358)
"If the test statistic's value is inconsistent with the null hypothesis, we reject the null hypothesis and infer that the alternative hypothesis is true." (Keller 2009, p. 348)
"Because the observed value ... is greater than the critical value ..., the decision is to reject the null hypothesis." (Black 2010, p. 359)
"If the obtained value is more extreme than the critical value, the null hypothesis cannot be accepted." (Salkind 2010, p. 243)
"The critical t-value ... must be surpassed by the observed t-value if the hypothesis test is to be statistically significant ...." (Zikmund and Babin 2010, p. 567)
"The calculated test statistic ... exceeds the upper boundary and falls into this rejection region. The null hypothesis is rejected." (Weiers 2011, p. 330)

You should note that all of the above quotes are used by statisticians and professors when discussing the results of an hypothesis test, and so you should not be surprised if someone asks you to summarize the results of a statistical test using a different language than the one we are using in this book.

3.4 End-of-Chapter Practice Problems

1. Suppose that you are an engineer working for a major tire manufacturer and that you have been asked to determine in a laboratory using specialized machines the "expected lifetime" of a new type of passenger sedan tire that has been made from a new type of synthetic material. Suppose, further, that your company would like to claim in its advertising for this new type of tire that this tire "lasts for 40,000 miles." You have been asked to take a random sample of tires produced with this new material, and to check their lifetime. You decide to test your Excel skills on a small number of tires, and the hypothetical data appear in Fig. 3.8:

EXPECTED LIFETIME OF A NEW TYPE OF PASSENGER CAR TIRE

Research question: "Does this new type of synthetic tire have an
                         expected lifetime of 40,000 miles?

LIFETIME IN MILES
38,400
39,500
39,400
42,300

46,700
45,800
44,300
38,600
42,500
41,600
40,200
38,600
37,900
38,900
40,600

Fig. 3.8 Worksheet Data for Chapter 3: Practice Problem #1

(a) To the right of this table, use Excel to find the sample size, mean, standard deviation, and standard error of the mean for the lifetime figures. Label your answers. Use number format (two decimal places) for the mean, standard deviation, and standard error of the mean.
(b) Enter the null hypothesis and the research hypothesis onto your spreadsheet.
(c) Use Excel's TINV function to find the 95 % confidence interval about the mean for these figures. Label your answers. Use number format (two decimal places).
(d) Enter your *result* onto your spreadsheet.
(e) Enter your *conclusion in plain English* onto your spreadsheet.
(f) Print the final spreadsheet to fit onto one page (if you need help remembering how to do this, see the objectives at the end of Chapter 2 in Sect. 2.4)
(g) On your printout, draw a diagram of this 95 % confidence interval by hand
(h) Save the file as: lifetime3

2. Suppose that you are an electrical engineer and that your company wants to determine if the lifetime of a new type of light bulb made with a new type of fusing process is 1300 hours under laboratory testing conditions. Let's suppose that you have taken a small random sample of light bulbs produced with this new process and that you want to analyze the hypothetical data in Fig. 3.9.

EXPECTED LIFETIME OF A NEW TYPE OF LIGHT BULB

Research question: "Does this new type of light bulb have an expected lifetime of 1300 hours?

LIFETIME IN HOURS

1252.3
1310.6
1264.1
1244.2

1282.8
1308.4
1319.4
1277.4
1289.7
1292.3
1256.2
1288.4
1279.9
1264.7
1305.8
1297.5

Fig. 3.9 Worksheet Data for Chapter 3: Practice Problem #2

Create an Excel spreadsheet with these data.

(a) Use Excel to the right of the table to find the sample size, mean, standard deviation, and standard error of the mean for these data. Label your answers, and use one decimal place for the mean, standard deviation, and standard error of the mean

(b) Enter the null hypothesis and the research hypothesis for this item on your spreadsheet.

(c) Use Excel's TINV function to find the 95 % confidence interval about the mean for these data. Label your answers on your spreadsheet. Use one decimal place for the lower limit and the upper limit of the confidence interval.

(d) Enter the *result* of the test on your spreadsheet.

(e) Enter the *conclusion* of the test in plain English on your spreadsheet.

(f) Print your final spreadsheet so that it fits onto one page (if you need help remembering how to do this, see the objectives at the end of Chapter 2 in Sect. 2.4).

(g) Draw a picture of the confidence interval, including the reference value, onto your spreadsheet.

(h) Save the final spreadsheet as: lightbulb3

3. Welch's sells a small can of what the company claims as "100 % Grape Juice" and the package states that the can contains 5.5 fluid ounces (FL.OZ.) of grape juice, and the can also labels this amount as 163 milliliters (ml) of grape juice. Suppose that you have been asked to take a random sample of cans produced today to see if they contain 163 ml of grape juice. You select a random sample of cans, measure the volume of liquid in them, and then start to analyze the hypothetical data are given in Fig. 3.10:

WELCH'S 100% GRAPE JUICE

Research question: "Does the average can of Welch's 100% grape juice
                        produced today contain 163 ml of grape juice?

ml
165
158
163
159
154
157
159
161
164
154
157
161
163

Fig. 3.10 Worksheet Data for Chapter 3: Practice Problem #3

Create an Excel spreadsheet with these data.

(a) Use Excel to the right of the table to find the sample size, mean, standard deviation, and standard error of the mean for these data. Label your answers, and use two decimal places for the mean, standard deviation, and standard error of the mean

(b) Enter the null hypothesis and the research hypothesis for this item onto your spreadsheet.

(c) Use Excel's TINV function to find the 95 % confidence interval about the mean for these data. Label your answers on your spreadsheet. Use two decimal places for the lower limit and the upper limit of the confidence interval.

(d) Enter the *result* of the test on your spreadsheet.

(e) Enter the *conclusion* of the test in plain English on your spreadsheet.

(f) Print your final spreadsheet so that it fits onto one page (if you need help remembering how to do this, see the objectives at the end of Chapter 2 in Sect. 2.4).

(g) Draw a picture of the confidence interval, including the reference value, onto your spreadsheet.

(h) Save the final spreadsheet as: grape3

References

Black K. Business statistics: for contemporary decision making. 6th ed. Hoboken: John Wiley & Sons, Inc.; 2010.

Keller G. Statistics for management and economics. 8th ed. Mason: South-Western Cengage Learning; 2009.

Ledolter R, Hogg R. Applied statistics for engineers and physical scientists. 3rd ed. Upper Saddle River: Pearson Prentice Hall; 2010.

McDaniel C, Gates R. Marketing research. 8th ed. Hoboken: John Wiley & Sons, Inc.; 2010.

McKillup S, Dyar M. Geostatistics explained: an introductory guide for earth scientists. Cambridge: Cambridge University Press; 2010.

Salkind N. Statistics for people who (think they) hate statistics. 2nd Excel 2007 ed. Los Angeles: Sage Publications; 2010.

Townend J. Practical statistics for environmental and biological scientists. Hoboken: John Wiley & Sons, Inc.; 2002.

Weiers R. Introduction to business statistics. 7th ed. Mason: South-Western Cengage Learning; 2011.

Zikmund W, Babin B. Exploring marketing research. 10th ed.. Mason: South- Western Cengage Learning; 2010.

Chapter 4
One-Group t-Test for the Mean

In this chapter, you will learn how to use one of the most popular and most helpful statistical tests in science research: the one-group t-test for the mean.

The formula for the one-group t-test is as follows:

$$t = \frac{\bar{X} - \mu}{S_{\bar{X}}} \quad \text{where} \tag{4.1}$$

$$\text{s.e.} = S_{\bar{X}} = \frac{s}{\sqrt{n}} \tag{4.2}$$

This formula asks you to take the mean $(\bar{X})$ and subtract the population mean (μ) from it, and then divide the answer by the standard error of the mean (s.e.). The standard error of the mean equals the standard deviation divided by the square root of n (the sample size). If you want to learn more about this test, see Schuenemeyer and Drew (2011).

Let's discuss the 7 STEPS of hypothesis testing using the one-group t-test so that you can understand how this test is used.

4.1 The 7 STEPS for Hypothesis-testing Using the One-group t-test

Objective: To learn the 7 steps of hypothesis-testing using the one-group t-test

Before you can try out your Excel skills on the one-group t-test, you need to learn the basic steps of hypothesis-testing for this statistical test. There are 7 steps in this process:

T.J. Quirk et al., *Excel 2010 for Physical Sciences Statistics: A Guide to Solving Practical Problems*, DOI 10.1007/978-3-319-00630-7_4,

4.1.1 STEP 1: State the null hypothesis and the research hypothesis

If you are using numerical scales in your survey, you need to remember that these hypotheses refer to the "middle" of the numerical scale. For example, if you are using 7-point scales with 1 = poor and 7 = excellent, these hypotheses would refer to the middle of these scales and would be:

Null hypothesis H_0: $\mu = 4$
Research hypothesis H_1: $\mu \neq 4$

As a second example, suppose that you worked for Honda Motor Company and that you wanted to place a magazine ad that claimed that the new Honda Fit got 35 miles per gallon (mpg). The hypotheses for testing this claim on actual data would be:

H_0 : $\mu = 35$ mpg
H_1 : $\mu \neq 35$ mpg

4.1.2 STEP 2: Select the appropriate statistical test

In this chapter we will be studying the one-group t-test, and so we will select that test.

4.1.3 STEP 3: Decide on a decision rule for the one-group t-test

(a) If the absolute value of t is less than the critical value of t, accept the null hypothesis.
(b) If the absolute value of t is greater than the critical value of t, reject the null hypothesis and accept the research hypothesis.

You are probably saying to yourself: "That sounds fine, but how do I find the absolute value of t?"

4.1.3.1 Finding the Absolute Value of a Number

To do that, we need another objective:

Objective: To find the absolute value of a number

If you took a basic algebra course in high school, you may remember the concept of "absolute value." In mathematical terms, the absolute value of any number is *always* that number expressed as a positive number.

For example, the absolute value of 2.35 is +2.35.

And the absolute value of minus 2.35 (i.e. −2.35) is also +2.35.

This becomes important when you are using the t-table in Appendix E of this book. We will discuss this table later when we get to Step 5 of the one-group t-test where we explain how to find the critical value of t using Appendix E.

4.1.4   STEP 4: Calculate the formula for the one-group t-test

Objective: To learn how to use the formula for the one-group t-test

The formula for the one-group t-test is as follows:

$$t = \frac{\bar{X} - \mu}{S_{\bar{X}}} \text{ where} \tag{4.1}$$

$$\text{s.e.} = S_{\bar{X}} = \frac{s}{\sqrt{n}} \tag{4.2}$$

This formula makes the following assumptions about the data (Foster et al. 1998): (1) The data are independent of each other (i.e., each person or event receives only one score), (2) the *population* of the data is normally distributed, and (3) the data have a constant variance (note that the standard deviation is the square root of the variance).

To use this formula, you need to follow these steps:

1. Take the sample mean in your research study and subtract the population mean μ from it (remember that the population mean for a study involving numerical rating scales is the "middle" number in the scale).
2. Then take your answer from the above step, and divide your answer by the standard error of the mean for your research study (you will remember that you learned how to find the standard error of the mean in Chapter 1; to find the standard error of the mean, just take the standard deviation of your research study and divide it by the square root of n, where n is the number of people or events used in your research study).
3. The number you get after you complete the above step is the value for t that results when you use the formula stated above.

4.1.5   STEP 5: Find the critical value of t in the t-table in Appendix E

Objective: To find the critical value of t in the t-table in Appendix E

Before we get into an explanation of what is meant by "the critical value of t," let's give you practice in finding the critical value of t by using the t-table in Appendix E.

Keep your finger on Appendix E as we explain how you need to "read" that table.

Since the test in this chapter is called the "one-group t-test," you will use the first column on the left in Appendix E to find the critical value of t for your research study (note that this column is headed: " n").

To find the critical value of t, you go down this first column until you find the sample size in your research study, and then you go to the right and read the critical value of t for that sample size in the critical t column in the table (note that *this is the column that you would use for both the one-group t-test and the 95 % confidence interval about the mean*).

For example, if you have 27 people in your research study, the critical value of t is 2.056.

If you have 38 people in your research study, the critical value of t is 2.026.

If you have more than 40 people in your research study, the critical value of t is always 1.96.

Note that the "critical t column" in Appendix E represents the value of t that you need to obtain to be 95 % confident of your results as "significant" results.

The critical value of t is the value that tells you whether or not you have found a "significant result" in your statistical test.

The t-table in Appendix E represents a series of "bell-shaped normal curves" (they are called bell-shaped because they look like the outline of the Liberty Bell that you can see in Philadelphia outside of Independence Hall).

The "middle" of these normal curves is treated as if it were zero point on the x-axis (the technical explanation of this fact is beyond the scope of this book, but any good statistics book (e.g. Zikmund and Babin 2010) will explain this concept to you if you are interested in learning more about it).

Thus, values of t that are to the right of this zero point are positive values that use a plus sign before them, and values of t that are to the left of this zero point are negative values that use a minus sign before them. Thus, some values of t are positive, and some are negative.

However, every statistics book that includes a t-table only reprints the *positive* side of the t-curves because the negative side is the mirror image of the positive side; this means that the negative side contains the exact same numbers as the positive side, but the negative numbers all have a minus sign in front of them.

Therefore, to use the t-table in Appendix E, you need to *take the absolute value of the t-value you found when you use the t-test formula* since the t-table in Appendix E only has the values of t that are the positive values for t.

Throughout this book, we are assuming that you want to be 95 % confident in the results of your statistical tests. Therefore, the value for t in the t-table in Appendix E tells you whether or not the t-value you obtained when you used the formula for the one-group t-test is within the 95 % interval of the t-curve range that that t-value would be expected to occur with 95 % confidence.

If the t-value you obtained when you used the formula for the one-group t-test is *inside* of the 95 % confidence range, we say that the result you found is *not significant* (note that this is equivalent to *accepting the null hypothesis!*).

If the t-value you found when you used the formula for the one-group t-test is *outside* of this 95 % confidence range, we say that you have found a *significant result* that would be expected to occur less than 5 % of the time (note that this is equivalent to *rejecting the null hypothesis and accepting the research hypothesis*).

4.1.6   STEP 6: State the result of your statistical test

There are two possible results when you use the one-group t-test, and only one of them can be accepted as "true."

Either: Since the absolute value of t that you found in the t-test formula is *less than the critical value of t* in Appendix E, you accept the null hypothesis.

Or: Since the absolute value of t that you found in the t-test formula is *greater than the critical value of t* in Appendix E, you reject the null hypothesis, and accept the research hypothesis.

4.1.7   STEP 7: State the conclusion of your statistical test in plain English!

In practice, this is more difficult than it sounds because you are trying to summarize the result of your statistical test in simple English that is both concise and accurate so that someone who has never had a statistics course (such as your boss, perhaps) can understand the result of your test. This is a difficult task, and we will give you lots of practice doing this last and most important step throughout this book.

If you have read this far, you are ready to sit down at your computer and perform the one-group t-test using Excel on some hypothetical data.

Let's give this a try.

4.2   One-group t-test for the mean

Suppose that you worked for a company that produces powdered graphite for use in gear lubrication of machines. Graphite as a lubrication in machines is particularly useful because it can be applied dry and does not attract soil or other material that may cause gears not to operate smoothly. Your company sells its powdered graphite in various containers and amounts. Other companies that purchase the graphite in "bulk" have traditionally been charged by volume or cubic meters (m^3). The company had several cubic meter containers that were filled with graphite and then transferred to plastic tubes for shipping. Suppose that your company wanted to determine if it was better to sell the graphite by mass (kg). Owners of the company believe that they are losing money because its bulk orders are measured by volume. Given the type of graphite that your company sells, there should be 650 kilograms (kg) of graphite per cubic meter. Your company conducts a series of tests where it measures a cubic meter of graphite and then measures the same graphite sample in terms of kilograms. You have been provided with a random sample of those tests.

Suppose further, that you have decided to analyze the data from the tests using the one-group t-test.

Suppose that the hypothetical data for these tests were based on a sample size of 124 samples which had a mean of 678 kg and a standard deviation of 144 kg.

Objective: To analyze the data using the one-group t-test

Create an Excel spreadsheet with the following information:

B11:   Null hypothesis:
B14:   Research hypothesis:

Note: In this situation, you know that one cubic meter of graphite should have 650 kg of graphite. Therefore, the hypotheses for this example are:

H_0 : $\mu = 650$ kg
H_1 : $\mu \neq 650$ kg

B17:   n
B20:   mean
B23:   STDEV
B26:   s.e.
B29:   critical t
B32:   t-test
B36:   Result:
B41:   Conclusion:

Now, use Excel:

D17:   enter the sample size
D20:   enter the mean
D23:   enter the STDEV (see Fig. 4.1)

Null hypothesis:	
Research hypothesis:	
n	124
mean	678
STDEV	144
s.e.	
critical t	
t-test	
Result:	
Conclusion:	

Fig. 4.1 Basic Data Table for Graphite (Practical Example)

D26   compute the standard error using the formula in Chapter 1
D29:   find the critical t value of t in the t-table in Appendix E

Now, enter the following formula in cell D32 to find the t-test result:

=(D20-650)/D26

This formula takes the sample mean (D20) and subtracts the population hypothesized mean of 650 from the sample mean, and THEN divides the answer by the standard error of the mean (D26). Note that you need to enter D20-650 with an open-parenthesis *before* D20 and a closed-parenthesis*after* 650 so that the *answer of 28 is THEN divided by the standard error of 12.93* to get the t-test result of 2.17.

Now, use two decimal places for both the s.e. and the t-test result (see Fig. 4.2).

Null hypothesis:	
Research hypothesis:	
n	124
mean	678
STDEV	144
s.e.	12.93
critical t	1.96
t-test	2.17
Result:	
Conclusion:	

Fig. 4.2   t-test Formula Result for Graphite Example

Now, write the following sentence in D36-D39 to summarize the result of the t-test:

D36:   Since the absolute value of t of 2.17 is
D37:   greater than the critical t of 1.96, we
D38:   reject the null hypothesis and accept
D39:   the research hypothesis.

Lastly, write the following sentence in D41-D44 to summarize the conclusion of the result for the graphite example:

D41:   There is significantly more than 650 kg per cubic meter
D42:   of graphite sold when measured by volume compared
D43:   to when the graphite is measured by mass, and it is
D44:   probably closer to 678 kg per cubic meter.

Save your file as: graphite4

Important note: We have used the term "significantly more" because the sample mean mass of 678 kg is greater than the hypothesized mean mass of 650 kg.

Important note: You are probably wondering why we entered both the result and the conclusion in separate cells instead of in just one cell. This is because if you enter them in one cell, you will be very disappointed when you print out your final spreadsheet, because one of two things will happen that you will not like: (1) if you print the spreadsheet to fit onto only one page, the result and the conclusion will force the entire spreadsheet to be printed in such small font size that you will be unable to read it, or (2) if you do not print the final spreadsheet to fit onto one page, both the result and the conclusion will "dribble over" onto a second page instead of fitting the entire spreadsheet onto one page. In either case, your spreadsheet will not have a "professional look."

Print the final spreadsheet so that it fits onto one page as given in Figure 4.3. Enter the null hypothesis and the research hypothesis by hand on your spreadsheet

Null hypothesis:	μ	=	650 kg
Research hypothesis:	μ	≠	650 kg
n	124		
mean	678		
STDEV	144		
s.e.	12.93		
critical t	1.96		
t-test	2.17		
Result:	Since the absolute value of t of 2.17 is greater than the critical t of 1.96, we reject the null hypothesis and accept the research hypothesis.		
Conclusion:	There is significantly more than 650 kg per cubic meter of graphite sold when measured by volume compared to when the graphite is measured by mass, and it is probably closer to 678 kg per cubic meter.		

Fig. 4.3 Final Spreadsheet for Graphite Example

Important Note: It is important for you to understand that "technically" the above conclusion in statistical terms should state:
"There is significantly more than 650 kg of graphite sold when measured by volume compared to when the graphite is measured by mass, and this result was probably not obtained by chance." However, throughout this book, we are using the term "significantly" in writing the conclusion of statistical tests to alert the reader that the result of the statistical test was probably not a

chance finding, but instead of writing all of those words each time, we use the word "significantly" as a shorthand to the longer explanation. This makes it much easier for the reader to understand the conclusion when it is written "in plain English," instead of technical, statistical language.

4.3   Can You Use Either the 95 Percent Confidence Interval About the Mean OR the One-Group t-test When Testing Hypotheses?

You are probably asking yourself:

"It sounds like you could use *either* the 95 % confidence interval about the mean *or* the one-groupt-test to analyze the results of the types of problems described so far in this book? Is this a correct statement?"

The answer is a resounding: *"Yes!"*

Both the confidence interval about the mean and the one-group t-test are used often in science research on the types of problems described so far in this book. *Both of these tests produce the same result and the same conclusion from the data set!*

Both of these tests are explained in this book because some researchers prefer the confidence interval about the mean test, others prefer the one-group t-test, and still others prefer to use both tests on the same data to make their results and conclusions clearer to the reader of their research reports. Since we do not know which of these tests your researcher prefers, we have explained both of them so that you are competent in the use of both tests in the analysis of statistical data.

Now, let's try your Excel skills on the one-group t-test on these three problems at the end of this chapter.

4.4   End-of-Chapter Practice Problems

1. Suppose that the U.S. Environmental Protection Agency (EPA) has set a maximum total phosphorus concentration (mg/L) for waste water effluent produced by chemical plants to be 0.015 mg/L. Suppose, further, that over a 90-day period, a random sample of waste water effluent was taken from a specific chemical plant and tested for phosphorus concentration. You have been asked to test your Excel skills on the hypothetical data given in Fig. 4.4.

PHOSPORUS CONCENTRATION (mg/L) IN WASTE WATER EFFLUENT

CONCENTRATION (mg/L)
0.0142
0.0135
0.0138
0.0136
0.0137
0.0135
0.0141
0.0140
0.0138
0.0134
0.0135
0.0137
0.0142
0.0132
0.0133

Fig. 4.4 Worksheet Data for Chapter 4: Practice Problem #1

(a) Write the null hypothesis and the research hypothesis on your spreadsheet
(b) Use Excel to find the sample size, mean, standard deviation, and standard error of the mean to the right of the data set. Use number format (4 decimal places) for the mean, standard deviation, and standard error of the mean.
(c) Enter the critical t from the t-table in Appendix E onto your spreadsheet, and label it.
(d) Use Excel to compute the t-value for these data (use 2 decimal places) and label it on your spreadsheet
(e) Type the result on your spreadsheet, and then type the conclusion in plain English on your spreadsheet
(f) Save the file as: waste31

2. Suppose that the Mayor of St. Louis wanted to reduce the amount of waste garbage collected in households in the Central West End of the city from an average of 26 kilograms (kg) per week in the weekly pickup cycle by instituting a recycling program for six months (26 weeks) in addition to the garbage collection program. You want to test your Excel skills to determine if there has been any change in the garbage collection program per household. The hypothetical data are presented in Fig. 4.5:

| WASTE GARBABE COLLECTED IN HOUSEHOLDS |

| Central West End, St. Louis, Missouri (USA) |

WEEKLY GARBAGE COLLECTION (kg)
18
21
20
19
21
23
22
24
18
17
19
20
21
24
26
21
18
25
19
26
21
23
22
24
26
19

Fig. 4.5 Worksheet Data for Chapter 4: Practice Problem #2

(a) *On your Excel spreadsheet*, write the null hypothesis and the research hypothesis for these data.

(b) Use Excel to find the sample size, mean, standard deviation, and standard error of the mean for these data (two decimal places for the mean, standard deviation, and standard error of the mean).

(c) Use Excel to perform a one-group t-test on these data (two decimal places).

(d) On your printout, type the critical value of t given in your t-table in Appendix E.

(e) On your spreadsheet, type the result of the t-test.

(f) On your spreadsheet, type the conclusion of your study in plain English.

(g) save the file as: garbage3

3. The state of Maine in the United States (USA) is famous for its lakes. There are more than 2,000 named lakes in Maine which is located in the northeastern seaboard of the USA. In addition, there are more than 4,000 other lakes in the state that are greater than one acre in size but have not been named. Dissolved oxygen (DO) is a measure of the quality of the water in a lake. The amount of DO declines as waste is entered into the lakes. Oxygen helps to break down the nutrients in the water, and Burt et al. (2009) state that the DO content of lakes should be 5 milligrams (mg) per liter (L). Suppose that you have collected data on a random sample of named lakes in Maine, and that you want to test your Excel skills on a small sample of these lakes before you try to analyze the data from a much larger sample. The hypothetical data appear in Fig. 4.6.

DISSOLVED OXYGEN CONTENT(DO) IN MAINE LAKES

DO (mg/L)
4.6
4.4
4.8
6.4
6.5
6.7
6.5
5.6
5.4
5.8
4.9
5.2
5.6
5.7
5.4
4.8

Fig. 4.6 Worksheet Data for Chapter 4: Practice problem #3

(a) Write the null hypothesis and the research hypothesis on your spreadsheet
(b) Use Excel to find the sample size, mean, standard deviation, and standard error of the mean to the right of the data set. Use number format (2 decimal places) for the mean, standard deviation, and standard error of the mean.

(c) Enter the critical t from the t-table in Appendix E onto your spreadsheet, and label it.

(d) Use Excel to compute the t-value for these data (use 2 decimal places) and label it on your spreadsheet

(e) Type the result on your spreadsheet, and then type the conclusion in plain English on your spreadsheet

(f) Save the file as: MElakes3

References

Burt J, Barber G, Rigby D. Elementary statistics for geographers. New York: The Guilford Press; 2009.

Foster D, Stine R, Waterman R. Basic business statistics: a casebook. New York: Springer-Verlag; 1998.

Schuenemeyer L, Drew L. Statistics for earth and environmental scientists. Hoboken: John Wiley & Sons; 2011.

Zikmund W, Babin B.Exploring marketing research.10th ed. Mason: South-Western Cengage Learning; 2010.

Chapter 5
Two-Group t-Test of the Difference of the Means for Independent Groups

Up until now in this book, you have been dealing with the situation in which you have had only one group of people or events in your research study and only one measurement "number" on each of these people or events. We will now change gears and deal with the situation in which you are measuring two groups instead of only one group.

Whenever you have two completely different groups of people or events (i.e., no one person or event is in both groups, but every person or event is measured on only one variable to produce one "number" for each person or event), we say that the two groups are "independent of one another." This chapter deals with just that situation and that is why it is called the two-group t-test for independent groups.

The two assumptions underlying the two-group t-test are the following (Wheater and Cook 2000): (1) both groups are sampled from a normal population, and (2) the variances of the two populations are approximately equal. Note that the standard deviation is merely the square root of the variance. (There are different formulas to use when each person or event is measured twice to create two groups of data, and this situation is called "dependent," but those formulas are beyond the scope of this book). This book only deals with two groups that are independent of one another so that no person or event is in both groups of data.

When you are testing for the difference between the means for two groups, it is important to remember that there are two different formulas that you need to use depending on the sample sizes of the two groups:

(1) Use Formula #1 in this chapter when both of the groups have a sample size greater than 30, and
(2) Use Formula #2 in this chapter when either one group, or both groups, have a sample size less than 30.

We will illustrate both of these situations in this chapter.

But, first, we need to understand the steps involved in hypothesis-testing when two groups are involved before we dive into the formulas for this test.

T.J. Quirk et al., *Excel 2010 for Physical Sciences Statistics: A Guide to Solving Practical Problems*, DOI 10.1007/978-3-319-00630-7_5,
© Springer International Publishing Switzerland 2013

5.1   The 9 STEPS for Hypothesis-testing Using the Two-group t-test

> Objective: To learn the 9 steps of hypothesis-testing using two groups of people or events and the two-group t-test

You will see that these steps parallel the steps used in the previous chapter that dealt with the one-group t-test, but there are some important differences between the steps that you need to understand clearly before we dive into the formulas for the two-group t-test.

5.1.1   STEP 1: Name one group, Group 1, and the other group, Group 2

The formulas used in this chapter will use the numbers 1 and 2 to distinguish between the two groups. If you define which group is Group 1 and which group is Group 2, you can use these numbers in your computations without having to write out the names of the groups.

For example, if you are were testing entering college freshmen who said that they wanted to major in Physics to see if there were gender differences in their SAT-Math scores as high school seniors, you could call the groups: "Freshmen Males" and "Freshmen Females," but this would require your writing out the words "Freshmen Males" and "Freshmen Females" whenever you wanted to refer to one of these groups. If you call the "Freshmen Males" group, Group 1, and the "Freshmen Females" group, Group 2, this makes it much easier to refer to the groups because it saves you writing time.

As a second example, you could be comparing the durability (i.e., lasting life) for two types of house paint, "latex paint" *versus* "oil-based paint." If you had to write out the names of the two types of paint whenever you wanted to refer to them, it would take you more time than it would if, instead, you named one type of paint, Group 1, and the other type of paint, Group 2.

Note, also, that it is completely arbitrary which group you call Group 1, and which Group you call Group 2. You will achieve the same result and the same conclusion from the formulas however you decide to define these two groups.

5.1.2   STEP 2: Create a table that summarizes the sample size, mean score, and standard deviation of each group

This step makes it easier for you to make sure that you are using the correct numbers in the formulas for the two-group t-test. If you get the numbers "mixed-up," your entire formula work will be incorrect and you will botch the problem terribly.

For example, suppose that you collected data on entering freshmen who said that they planned to become Physics majors and found that the Freshmen Males group had 57 men in it and their SAT-Math scores averaged 610 with a standard deviation of 120, while the Freshmen Females group had 46 females in it and their SAT-Math scores averaged 640 with a standard deviation of 110.

The formulas for analyzing these data to determine if there was a significant different in the average SAT-Math score for Freshmen Males *versus* Freshmen Females require you to use six numbers correctly in the formulas: the sample size, the mean, and the standard deviation of each of the two groups. All six of these numbers must be used correctly in the formulas if you are to analyze the data correctly.

If you create a table to summarize these data, a good example of the table, using both Step 1 and Step 2, would be the data presented in Fig. 5.1:

	A	B	C	D	E	F
1						
2						
3		Group	n	Mean	STDEV	
4		1 (name it)				
5		2 (name it)				
6						
7						

Fig. 5.1 Basic Table Format for the Two-group t-test

For example, if you decide to call Group 1 the Freshmen Males group and Group 2 the Freshmen Females group, the following table would place the six numbers from your research study into the proper cells of the table as in Fig. 5.2:

	A	B	C	D	E	
1						
2						
3		Group	n	Mean	STDEV	
4		1 Freshmen Males SAT-Math scores	57	610	120	
5		2 Freshmen Females SAT-Math scores	46	640	110	
6						
7						

Fig. 5.2 Results of Entering the Data Needed for the Two-group t-test

You can now use the formulas for the two-group t-test with more confidence that the six numbers will be placed in the proper place in the formulas.

Note that you could just as easily call Group 1 the Freshmen Females group and Group 2 the Freshmen Males group; it makes no difference how you decide to name the two groups; this decision is up to you and you will get the same result to your statistical test no matter which decision you make.

5.1.3   STEP 3: State the null hypothesis and the research hypothesis for the two-group t-test

If you have completed Step 1 above, this step is very easy because the null hypothesis and the research hypothesis will always be stated in the same way for the two-group t-test. The null hypothesis states that the population means of the two groups are equal, while the research hypothesis states that the population means of the two groups are not equal. In notation format, this becomes:

H_0 : $\mu_1 = \mu_2$
H_1 : $\mu_1 \neq \mu_2$

You can now see that this notation is much simpler than having to write out the names of the two groups in all of your formulas.

5.1.4   STEP 4: Select the appropriate statistical test

Since this chapter deals with the situation in which you have two groups but only one measurement on each person or event in each group, we will use the two-group t-test throughout this chapter.

5.1.5   STEP 5: Decide on a decision rule for the two-group t-test

The decision rule is exactly what it was in the previous chapter (see Section 4.1.3) when we dealt with the one-group t-test.

(a) If the absolute value of t is less than the critical value of t, accept the null hypothesis.
(b) If the absolute value of t is greater than the critical value of t, reject the null hypothesis and accept the research hypothesis.

Since you learned how to find the absolute value of t in the previous chapter (see Sect. 4.1.3.1), you can use that knowledge in this chapter.

5.1.6   STEP 6: Calculate the formula for the two-group t-test

Since we are using two different formulas in this chapter for the two-group t-test depending on the sample size in the two groups, we will explain how to use those formulas later in this chapter.

5.1.7 STEP 7: Find the critical value of t in the t-table in Appendix E

In the previous chapter where we were dealing with the one-group t-test, you found the critical value of t in the t-table in Appendix E by finding the sample size for the one group in the first column of the table, and then reading the critical value of t across from it on the right in the "critical t column" in the table (see Section 4.1.5). This process was fairly simple once you have had some practice in doing this step.

However, for the two-group t-test, the procedure for finding the critical value of t is more complicated because you have two different groups in your study, and they often have different sample sizes in each group.

To use Appendix E correctly in this chapter, you need to learn how to find the "degrees of freedom" for your study. We will discuss that process now.

5.1.7.1 Finding the degrees of freedom (df) for the Two-group t-test

> Objective: To find the degrees of freedom for the two-group t-test and to use it to find the critical value of t in the t-table in Appendix E

The mathematical explanation of the concept of the "degrees of freedom" is beyond the scope of this book, but you can find out more about this concept by reading any good statistics book (e.g. Keller, 2009). For our purposes, you can easily understand how to find the degrees of freedom and to use it to find the critical value of t in Appendix E. The formula for the degrees of freedom (df) is:

$$\text{degrees of freedom} = df = n_1 + n_2 - 2 \qquad (5.1)$$

In other words, you add the sample size for Group 1 to the sample size for Group 2 and then subtract 2 from this total to get the number of degrees of freedom to use in Appendix E.

Take a look at Appendix E.

Instead of using the first column as we did in the one-group t-test that is based on the sample size, n , of one group, we need to use the second-column of this table (df) to find the critical value of t for the two-group t-test.

For example, if you had 13 people in Group 1 and 17 people in Group 2, the degrees of freedom would be: $13 + 17 - 2 = 28$, and the critical value of t would be 2.048 *since you look down the second column which contains the degrees of freedom* until you come to the number 28, and then read 2.048 in the "critical t column" in the table to find the critical value of t when $df = 28$.

As a second example, if you had 52 people in Group 1 and 57 people in Group 2, the degrees of freedom would be: $52 + 57 - 2 = 107$. When you go down the second column in Appendix E for the degrees of freedom, you find that *once you go beyond the degrees of freedom equal to 39, the critical value of t is always 1.96,* and that is the value you would use for the critical t with this example.

5.1.8   STEP 8: State the result of your statistical test

The result follows the exact same result format that you found for the one-group t-test in the previous chapter (see Section 4.1.6):

Either: Since the absolute value of t that you found in the t-test formula is *less than the critical value of t* in Appendix E, you accept the null hypothesis.
Or: Since the absolute value of t that you found in the t-test formula is *greater than the critical value of t* in Appendix E, you reject the null hypothesis and accept the research hypothesis.

5.1.9   STEP 9: State the conclusion of your statistical test in plain English!

Writing the conclusion for the two-group t-test is more difficult than writing the conclusion for the one-group t-test because you have to decide what the difference was between the two groups.

When you accept the null hypothesis, the conclusion is simple to write: "There is no difference between the two groups in the variable that was measured."

But when you reject the null hypothesis and accept the research hypothesis, you need to be careful about writing the conclusion so that it is both accurate and concise.

Let's give you some practice in writing the conclusion of a two-group t-test.

5.1.9.1   Writing the Conclusion of the Two-group t-test When You Accept the Null Hypothesis

Objective: To write the conclusion of the two-group t-test when you have accepted the null hypothesis

Suppose that you worked for a company that produces engineered (man-made) stone. The most popular stone that your company produces has always had complaints about being too soft. Generally, the hardness of stone is measured using the Mohs scale of hardness. The Mohs scale of hardness ranges from 1 to 10 with one being the softest (talc) and 10 being the hardest (diamond). Your company has changed the chemical formula for their most popular stone to try to increase its hardness. A test has been developed using the Mohs scale, and tried out with a pilot study with just a few samples to see how it was working. Item #10 of this survey is given in Fig. 5.3.

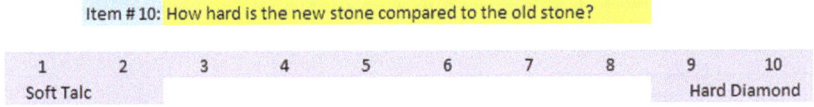

| Item # 10: How hard is the new stone compared to the old stone? |

| 1 | 2 | 3 | 4 | 5 | 6 | 7 | 8 | 9 | 10 |
| Soft Talc | | | | | | | | Hard Diamond | |

Fig. 5.3 Mohs Scale Survey Item #10

Suppose further, that you have decided to analyze the data from the tests comparing the "Old Stone" to the "New Stone" by using the two-group t-test.

Important note: You would need to use this test for each of the survey items separately.

Suppose that the hypothetical data for Item #10 was based on a sample size of 124 pieces of "Old Stone" which had a mean score on this item of 6.58 and a standard deviation on this item of 2.44. Suppose that you also had data from 86 samples of the "New Stone" which had a mean score of 6.45 with a standard deviation of 1.86.

We will explain later in this chapter how to produce the results of the two-group t-test using its formulas, but, for now, let's "cut to the chase" and tell you that those formulas would produce the following in Fig. 5.4:

	A	B	C	D	E	F
1						
2						
3		Group	n	Mean	STDEV	
4		1 Old Stone	124	6.58	2.44	
5		2 New Stone	86	6.45	1.86	
6						

Fig. 5.4 Worksheet Data for the Mohs Scale for Item #10 for Accepting the Null Hypothesis

degrees of freedom:	208
critical t:	1.96 (in Appendix E)
t-test formula:	0.44 (when you use your calculator!)
Result:	Since the absolute value of 0.44 is less than the critical t of 1.96, we accept the null hypothesis.
Conclusion:	There was no difference between the "Old Stone" and the "New Stone" in their hardness using the Mohs scale of hardness.

Now, let's see what happens when you reject the null hypothesis (H_0) and accept the research hypothesis (H_1).

5.1.9.2 Writing the Conclusion of the Two-group t-test When You Reject the Null Hypothesis and Accept the Research Hypothesis

> Objective: To write the conclusion of the two-group t-test when you have rejected the null hypothesis and accepted the research hypothesis

Let's continue with this same example, but with the result that we reject the null hypothesis and accept the research hypothesis.

Let's assume that this time you have "Old Stone" data on 85 samples and their mean score on this question was 7.26 with a standard deviation of 2.35. Let's further suppose that you also have data on 48 "New Stone" samples and their mean score on this question was 4.37 with a standard deviation of 3.26.

Without going into the details of the formulas for the two-group t-test, these data would produce the following result and conclusion based on Fig. 5.5:

	A	B	C	D	E	F
1						
2						
3		Group	n	Mean	STDEV	
4		1 Old Stone	85	7.26	2.35	
5		2 New Stone	48	4.37	3.26	
6						

Fig. 5.5 Worksheet Data for Item #10 for Obtaining a Significant Difference between the Two Types of Stone

Null Hypothesis:	$\mu_1 = \mu_2$
Research Hypothesis:	$\mu_1 \neq \mu_2$
degrees of freedom:	131
critical t:	1.96 (in Appendix E)
t-test formula:	5.40 (when you use your calculator!)
Result:	Since the absolute value of 5.40 is greater than the critical t of 1.96, we reject the null hypothesis and accept the research hypothesis.

Now, you need to compare the ratings on the old stone and the new stone to find out which type of stonehad a harder rating according to the Mohs hardness scale using the following rule:

Rule: To summarize the conclusion of the two-group t-test, just compare the means of the two groups, and be sure to use the word "significantly" in your conclusion if you rejected the null hypothesis and accepted the research hypothesis.

A good way to prepare to write the conclusion of the two-group t-test when you are using a rating scale is to place the mean scores of the two groups on a drawing of the scale so that you can visualize the difference of the mean scores. For example, for our stone hardness example above, you would draw this "picture" of the scale in Fig. 5.6:

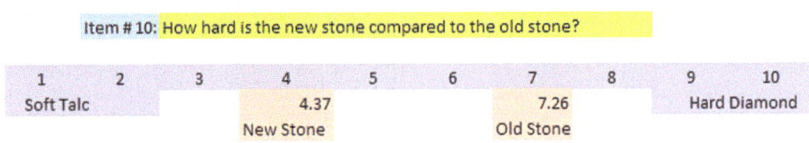

Fig. 5.6 Example of Drawing a "Picture" of the Means of the Two Groups on the Rating Scale

This drawing tells you visually that the Old Stone was harder than the New Stone on this item (7.26 vs. 4.37). *And, since you rejected the null hypothesis and accepted the research hypothesis, you know that you have found a significant difference between the two mean scores.*

So, our conclusion needs to contain the following key words:

– Old Stone
– New Stone
– Mohs hardness scale
– significantly
– harder or softer
– *either*(7.26 vs. 4.37)*or*(4.37 vs. 7.26)

We can use these key words to write the either of two conclusions which are *logically identical*:

Either: The Old Stones were significantly harder than the New Stones according to the Mohs scale of hardness (7.26 vs. 4.37).

Or: The New Stones were significantly softer than the Old Stones according to the Mohs scale of hardness (4.37 vs. 7.26).

Both of these conclusions are accurate, so you can decide which one you want to write. It is your choice.

Also, note that the mean scores in parentheses at the end of these conclusions must match the sequence of the two groups in your conclusion. For example, if you say that: "The Old stones were significantly harder than the New Stones according to the Mohs scale of hardness," the end of this conclusion should be: (7.26 vs. 4.37) since you mentioned Old Stones first, and New Stones second.

Alternately, if you wrote that: "The New Stones were significantly softer than the Old Stones according to the Mohs scale of hardness," the end of this conclusion should be: (4.37 vs. 7.26) since you mentioned New Stones first, and Old Stones second.

Putting the two mean scores at the end of your conclusion saves the reader from having to turn back to the table in your research report to find these mean scores to see how far apart the mean scores were.

Now, let's discuss FORMULA #1 that deals with the situation in which both groups have a sample size greater than 30.

Objective: To use FORMULA #1 for the two-group t-test when both groups
         have a sample size greater than 30

5.2 FORMULA #1: Both Groups Have a Sample Size Greater Than 30

The first formula we will discuss will be used when you have two groups with a sample size greater than 30 in each group and one measurement on each member in each group. This formula for the two-group t-test is:

$$t = \frac{\overline{X}_1 - \overline{X}_2}{S_{\overline{X}_1 - \overline{X}_2}} \tag{5.2}$$

where

$$S_{\overline{X}_1 - \overline{X}_2} = \sqrt{\frac{S_1^2}{n_1} + \frac{S_2^2}{n_2}} \tag{5.3}$$

and where degrees of freedom $= df = n_1 + n_2 - 2$ (5.1)

This formula looks daunting when you first see it, but let's explain some of the parts of this formula:

We have explained the concept of "degrees of freedom" earlier in this chapter, and so you should be able to find the degrees of freedom needed for this formula in order to find the critical value of t in Appendix E.

In the previous chapter, *the formula for the one-group t-test was the following*:

$$t = \frac{\overline{X} - \mu}{S_{\overline{X}}} \qquad (4.1)$$

$$\text{where s.e.} = S_{\overline{X}} = \frac{S}{\sqrt{n}} \qquad (4.2)$$

For the one-group t-test, you found the mean score and subtracted the population mean from it, and then divided the result by the standard error of the mean (s.e.) to get the result of the t-test. You then compared the t-test result to the critical value of t to see if you either accepted the null hypothesis, or rejected the null hypothesis and accepted the research hypothesis.

The two-group t-test requires a different formula because you have two groups, each with a mean score on some variable. You are trying to determine whether to accept the null hypothesis that the *population means of the two groups are equal* (in other words, there is no difference statistically between these two means), or whether the difference between the means of the two groups is "sufficiently large" that you would accept *that there is a significant difference* in the mean scores of the two groups.

The numerator of the two-group t-test asks you to find the difference of the means of the two groups:

$$\overline{X}_1 - \overline{X}_2 \qquad (5.4)$$

The next step in the formula for the two-group t-test is to divide the answer you get when you subtract the two means by the standard error of the difference of the two means, and *this is a different standard error of the mean that you found for the one-group t-test because there are two means in the two-group t-test.*

The standard error of the mean when you have two groups is called the "standard error of the difference of the means." This formula looks less scary when you break it down into four steps:

1. Square the standard deviation of Group 1, and divide this result by the sample size for Group 1 (n_1).
2. Square the standard deviation of Group 2, and divide this result by the sample size for Group 2 (n_2).
3. Add the results of the above two steps to get a total score.
4. *Take the square root of this total score* to find the standard error of the difference of the means between the two groups, $S_{\overline{X}_1 - \overline{X}_2} = \sqrt{\frac{S_1^2}{n_1} + \frac{S_2^2}{n_2}}$

This last step is the one that gives students the most difficulty when they are finding this standard error using their calculator, because they are in such a hurry to get to the answer that they forget to carry the square root sign down to the last step, and thus get a larger number than they should for the standard error.

5.2.1   An example of Formula #1 for the Two-group t-test

Now, let's use Formula #1 in a situation in which both groups have a sample size greater than 30.

Suppose that a large manufacturing company produced several types of thermoses using two different metal alloys (A and B) to form the thermos bodies, and that it wanted to compare these alloys to see if there were differences in their insulating properties for the thermoses. The company has conducted a test in which 100 °C water was placed in a thermos with a temperature probe and the lid was then closed on the container. The temperature of the water was taken 8 hours later. Suppose, further, that the results of this test are given in Fig. 5.7.

Insulating properties:	Temperature (° C) of water inside the thermos after 8 hours									
0	10	20	30	40	50	60	70	80	90	100
Frozen										Boiling

Fig. 5.7 Example of a Temperature Scale Rating for Water Temperature Inside the Thermos (Practical Example)

Suppose you collect these measurements and determine (using your new Excel skills) that the 52 thermoses in Group A had a mean temperature of 55 with a standard deviation of 7, while the 57 thermoses in Group B had a mean temperature of 64 with a standard deviation of 13.

Note that the two-group t-test does not require that both groups have the same sample size. This is another way of saying that the two-group t-test is "robust" (a fancy term that statisticians like to use).

Your data then produce the following table in Fig. 5.8:

	A	B	C	D	E	F
1						
2						
3		Group	n	Mean	STDEV	
4		1 Group A	52	55	7	
5		2 Group B	57	64	13	
6						

Fig. 5.8 Worksheet Data for Water Temperature

Create an Excel spreadsheet, and enter the following information:

B3: Group
B4: 1 Group A
B5: 2 Group B
C3: n
D3: Mean
E3: STDEV
C4: 52
D4: 55
E4: 7
C5: 57
D5: 64
E5: 13

Now, widen column B so that it is twice as wide as column A, and center the six numbers and their labels in your table (see Fig. 5.9)

	A	B	C	D	E	F
1						
2						
3		Group	n	Mean	STDEV	
4		1 Group A	52	55	7	
5		2 Group B	57	64	13	
6						

Fig. 5.9 Results of Widening Column B and Centering the Numbers in the Cells

B8: Null hypothesis:
B10: Research hypothesis:

Since both groups have a sample size greater than 30, you need to use Formula #1 for the t-test for the difference of the means of the two groups.

Let's "break this formula down into pieces" to reduce the chance of making a mistake.

B13: STDEV1 squared / n1(note that you square the standard deviation of Group 1, and then divide the result by the sample size of Group 1)
B16: STDEV2 squared / n2
B19: D13 + D16
B22: s.e.

B25:   critical t
B28:   t-test
B31:   Result:
B36:   Conclusion: (see Fig. 5.10)

Fig. 5.10 Formula Labels
for the Two-group t-test

Group	n	Mean	STDEV
1 Group A	52	55	7
2 Group B	57	64	13

Null hypothesis:

Research hypothesis:

STDEV1 squared/n1

STDEV2 squared/n2

D13 + D16

s.e.

critical t

t-test

Result:

Conclusion:

You now need to compute the values of the above formulas in the following cells:

D13:   the result of the formula needed to compute cell B13 (use 2 decimals)
D16:   the result of the formula needed to compute cell B16 (use 2 decimals)
D19:   the result of the formula needed to compute cell B19 (use 2 decimals)
D22:   =SQRT(D19)   (use 2 decimals)

This formula should give you a standard error (s.e.) of 1.98.

D25:   1.96
        (Since df = n1 + n2 – 2, this gives df = 109 – 2 = 107, and the
        critical t is, therefore, 1.96 in Appendix E.)
D28:   =(D4–D5)/D22   (use 2 decimals)

This formula should give you a value for the t-test of: −4.55.

Next, check to see if you have rounded off all figures in D13: D28 to two decimal places (see Fig. 5.11).

Fig. 5.11 Results of the
t-test Formula for Water
Temperature Comparisons

Group	n	Mean	STDEV
1 Group A	52	55	7
2 Group B	57	64	13

Null hypothesis:	
Research hypothesis:	
STDEV1 squared/n1	0.94
STDEV2 squared/n2	2.96
D13 + D16	3.91
s.e.	1.98
critical t	1.96
t-test	−4.55
Result:	
Conclusion:	

Now, write the following sentence in D31 to D34 to summarize the result of the study:

D31:   Since the absolute value of − 4.55
D32:   is greater than the critical t of
D33:   1.96, we reject the null hypothesis
D34:   and accept the research hypothesis.

Finally, write the following sentence in D36 to D38 to summarize the conclusion of the study in plain English:

D36:   Overall, Thermoses from Group B were significantly
D37:   better at insulating than Thermoses from Group A
D38:   (64 vs. 55).

Save your file as: TEMP12E

Important note: You are probably wondering why we entered both the result and the conclusion in separate cells instead of in just one cell. This is because if you enter them in one cell, you will be very disappointed when you print out your final spreadsheet, because one of two things will happen that you will not like: (1) if you print the spreadsheet to fit onto only one page, the result and the conclusion will force the entire spreadsheet to be printed in such small font size that you will be unable to read it, or (2) if you do not print the final spreadsheet to fit onto one page, both the result and the conclusion will "dribble over" onto a second page instead of fitting the entire spreadsheet onto one page. In either case, your spreadsheet will not have a "professional look."

Print this file so that it fits onto one page, and write by hand the null hypothesis and the research hypothesis on your printout.

The final spreadsheet appears in Figure 5.12.

Group	n	Mean	STDEV
1 Group A	52	55	7
2 Group B	57	64	13

Null hypothesis:		μ_1	=	μ_2

Research hypothesis:		μ_1	≠	μ_2

STDEV1 squared/n1	0.94
STDEV2 squared/n2	2.96
D13 + D16	3.91
s.e.	1.98
critical t	1.96
t-test	− 4.55

Result:	Since the absolute value of − 4.55 is greater than the critical t of 1.96, we reject the null hypothesis and accept the research hypothesis.
Conclusion:	Overall, Thermoses from Group B were significantly better at insulating than Thermoses from Group A (64 vs. 55)

Fig. 5.12 Final Worksheet for Water Temperature Comparisons

Now, let's use the second formula for the two-group t-test which we use whenever either one group, or both groups, have a sample size less than 30.

Objective: To use Formula #2 for the two-group t-test when one or both groups
          have a sample size less than 30

Now, let's look at the case when one or both groups have a sample size less than 30.

5.3   FORMULA #2: One or Both Groups Have a Sample Size Less Than 30

Suppose that you are an electrical engineer and you have been asked to compare the number of hours until failure of two new models of light bulbs (Model A and Model B) that have been prepared by your company's Research & Development department. Suppose, further, that you have decided to analyze the data from this study using the two-group t-test for independent samples. You decide to try out your new Excel skills on a small sample of light bulbs of each model on the hypothetical data given in Fig. 5.13:

Fig. 5.13 Worksheet Data
for Light Bulbs (Practical
Example)

LIGHT BULB HOURS (hrs) UNTIL FAILURE

Model A	Model B
910	890
940	940
980	950
1005	960
842	913
836	908
869	1030
910	1050
930	1040
897	
864	

Let's call Model A as Group 1, and Model B as Group 2.

Null hypothesis: $\mu_1 = \mu_2$
Research hypothesis: $\mu_1 \neq \mu_2$

Note: Since both groups have a sample size less than 30, you need to use Formula #2 in the following steps:

Create an Excel spreadsheet, and enter the following information:

B2:   LIGHT BULB HOURS (hrs) UNTIL FAILURE
B4:   Model A
C4:   Model B
B5:   910
B15:   864
C5:   890
C13:   1040

Now, enter the other figures into this table. Be sure to double-check all of your figures. If you have only one incorrect figure, you will not be able to obtain the correct answer to this problem.

Now, widen columns B and C so that all of the information fits inside the cells. To do this, click on both letters B and C at the top of these columns on your spreadsheet to highlight all of the cells in columns B and C. Then, move the mouse pointer to the right end of the B cell until you get a "cross" sign; then, click on this cross sign and drag the sign to the right until you can read all of the words on your screen. Then, stop clicking! Both Column B and Column C should now be the same width.

Then, center all information in the table except the top title by using the following steps:

Left-click your mouse and highlight cells B4:C15. Then, click on the bottom line, second from the left icon, under "Alignment" at the top-center of Home. All of the information in the table should now be in the center of each cell.

E5:   Null hypothesis:
E7:   Research hypothesis:
E9:   Group
E10:   1 Model A
E11:   2 Model B
F9:   n
G9:   Mean
H9:   STDEV

Your spreadsheet should now look like Fig. 5.14.

LIGHT BULB HOURS (hrs) UNTIL FAILURE

Model A	Model B
910	890
940	940
980	950
1005	960
842	913
836	908
869	1030
910	1050
930	1040
897	
864	

Null hypothesis:

Research hypothesis:

Group	n	Mean	STDEV
1 Model A			
2 Model B			

Fig. 5.14 Light Bulb Hours Until Failure Worksheet Data for Hypothesis Testing

Now you need to use your Excel skills from Chapter 1 to fill in the sample sizes (n), the Means, and the Standard Deviations (STDEV) in the Table in cells F10:H11. Be sure to double-check your work or you will not be able to obtain the correct answer to this problem if you have only one incorrect figure! Round off the means and standard deviations to zero decimal places and center these six figures within their cells.

Since both groups have a sample size less than 30, you need to use Formula #2 for the t-test for the difference of the means of two independent samples.

Formula #2 for the two-group t-test is the following:

$$t = \frac{\overline{X}_1 - \overline{X}_2}{S_{\overline{X}_1 - \overline{X}_2}} \tag{5.1}$$

where

$$S_{\overline{X}_1 - \overline{X}_2} = \sqrt{\frac{(n_1 - 1)S_1^2 + (n_2 - 1)S_2^2}{n_1 + n_2 - 2}\left(\frac{1}{n_1} + \frac{1}{n_2}\right)} \tag{5.5}$$

and where degrees of freedom $= df = n_1 + n_2 - 2$ $\qquad$ (5.6)

This formula is complicated, and so it will reduce your chance of making a mistake in writing it if you "break it down into pieces" instead of trying to write the formula as one cell entry.

Now, enter these words on your spreadsheet:

E14:   (n1 − 1) x STDEV1 squared
E16:   (n2 − 1) x STDEV2 squared

E18:　　$n_1 + n_2 - 2$
E20:　　$1/n_1 + 1/n_2$
E23:　　s.e.
E26:　　critical t
E29:　　t-test
B32:　　Result:
B36:　　Conclusion: (see Fig. 5.15)

LIGHT BULB HOURS (hrs) UNTIL FAILURE

Model A	Model B
910	890
940	940
980	950
1005	960
842	913
836	908
869	1030
910	1050
930	1040
897	
864	

Null hypothesis:

Research hypothesis:

Group	n	Mean	STDEV
1 Model A	11	908	54
2 Model B	9	965	61

(n1 - 1) x STDEV1 squared

(n2-1) x STDEV2 squared

n1 + n2 - 2

1/n1 + 1/n2

s.e.

critical t

t-test

Result:

Conclusion:

Fig. 5.15 Light Bulb Formula Labels for the Two-group t-test

You now need to use your Excel skills to compute the values of the above formulas in the following cells:

H14:   the result of the formula needed to compute cell E14 (use 2 decimals)
H16:   the result of the formula needed to compute cell E16 (use 2 decimals)
H18:   the result of the formula needed to compute cell E18
H20:   the result of the formula needed to compute cell E20 (use 2 decimals)
H28:   =SQRT(((H14 + H16)/H18)*H20)

*Note the three open-parentheses after SQRT, and the three closed parentheses on the right side of this formula.*You need three open parentheses and three closed parentheses in this formula or the formula will not work correctly.

The above formula gives a standard error of the difference of the means equal to 25.68(two decimals) in cell H23.

H26:   Enter the critical t value from the t-table in Appendix E in this cell using
       $df = n_1 + n_2 - 2$ to find the critical t value
H29:   =(G10-G11)/H23

Note that you need an open-parenthesis *before G10* and a closed-parenthesis *after G11* so that this answer of -57 is *THEN* divided by the standard error of the difference of the means of 25.68, to give a t-test value of -2.22. Use two decimal places for the t-test result (see Fig. 5.16).

LIGHT BULB HOURS (hrs) UNTIL FAILURE

Model A	Model B
910	890
940	940
980	950
1005	960
842	913
836	908
869	1030
910	1050
930	1040
897	
864	

Null hypothesis:

Research hypothesis:

Group	n	Mean	STDEV
1 Model A	11	908	54
2 Model B	9	965	61

(n1 - 1) x STDEV1 squared	29224.73
(n2-1) x STDEV2 squared	29526.22
n1 + n2 - 2	18
1/n1 + 1/n2	0.20
s.e.	25.68
critical t	2.101
t-test	-2.22

Result:

Conclusion:

Fig. 5.16 Light Bulb Hours Two-group t-test Formula Results

Now write the following sentence in C32 to C33 to summarize the *result* of the study:

C32:   Since the absolute value of -2.22 is greater than 2.101,
C33:   we reject the null hypothesis and accept the research hypothesis.

Finally, write the following sentence in C36 to C37 to summarize the *conclusion* of the study:

C36:   Model B lasted significantly more hours until failure than Model A
C37:   (965 hours vs. 908 hours).

Save your file as: bulb3

Print the final spreadsheet so that it fits onto one page.
Write the null hypothesis and the research hypothesis by hand on your printout.
The final spreadsheet appears in Figure 5.17.

LIGHT BULB HOURS (hrs) UNTIL FAILURE

Model A	Model B
910	890
940	940
980	950
1005	960
842	913
836	908
869	1030
910	1050
930	1040
897	
864	

Null hypothesis: μ_1 = μ_2

Research hypothesis: μ_1 ≠ μ_2

Group	n	Mean	STDEV
1 Model A	11	908	54
2 Model B	9	965	61

(n1 - 1) x STDEV1 squared	29224.73
(n2-1) x STDEV2 squared	29526.22
n1 + n2 - 2	18
1/n1 + 1/n2	0.20
s.e.	25.68
critical t	2.101
t-test	-2.22

Result: Since the absolute value of − 2.22 is greater than the critical t of 2.101,
             we reject the null hypothesis and accept the research hypothesis.

Conclusion: Model B lasted significantly more hours until failure than Model A
             (965 hours vs. 908 hours).

Fig. 5.17 Light Bulb Hours Final Spreadsheet

5.4   End-of-Chapter Practice Problems

1. Suppose that you are an electrical engineer and that you have been asked to perform a laboratory test on two types of lead wires that can be used in the manufacturing of household light bulbs. In each production run of the wire currently being used by your company, some of the wires did not feed properly into the machines. Your company is experimenting with a new type of lead wire to be used in production if the average number of misfeeding lead wires per hour for the New Wire is less than the number of misfeeding lead wires per hour for the Current Wire being used in production. 112 Current Wires were tested and had an average of 21.1 misfeeds per hour with a standard deviation of 3.24 misfeeds per hour. 126 New Wires were tested and had an average of 19.6 misfeeds per hour with a standard deviation of 3.06 misfeeds per hour.

   (a) State the null hypothesis and the research hypothesis on an Excel spreadsheet.
   (b) Find the standard error of the difference between the means using Excel
   (c) Find the critical t value using Appendix E, and enter it on your spreadsheet.
   (d) Perform a t-test on these data using Excel. What is the value of t that you obtain?
   Use three decimal places for all figures in the formula section of your spreadsheet.
   (e) State your result on your spreadsheet.
   (f) State your conclusion in plain English on your spreadsheet.
   (g) Save the file as: leadwire3

2. Suppose that you have been hired to analyze the data for a research project in which two types of household primer paint were tested to determine their average drying time. The current paint used by your company is being compared to a new paint that has been introduced into the market and your company wants to know if these paints differ in their average drying time in minutes. An experiment has been conducted, and the data have been turned over to you. You want to test your Excel skills on the hypothetical data given in Fig. 5.18.

Fig. 5.18 Worksheet Data
for Chapter 5: Practice
Problem #2

DRYING TIME FOR HOUSEHOLD PRIMER PAINT (minutes)

Current paint	New paint
118	111
116	115
121	112
124	113
117	114
115	119
110	118
117	117
119	120
121	114
123	112
125	

(a) On your Excel spreadsheet, write the null hypothesis and the research hypothesis.
(b) Create a table that summarizes these data on your spreadsheet and use Excel to find the sample sizes, the means, and the standard deviations of the two types of paint in this table. Use two decimals for the means and standard deviations.
(c) Use Excel to find the standard error of the difference of the means (two decimal places).
(d) Use Excel to perform a two-group t-test. What is the value of t that you obtain (use two decimal places)?
(e) On your spreadsheet, type the critical value of t using the t-table in Appendix E.
(f) Type your result on the test on your spreadsheet.
(g) Type your conclusion in plain English on your spreadsheet.
(h) save the file as: PAINT3

3. Suppose that the CEO of a national fishing hook manufacturing company has asked you to "run the data" to see if there is a difference in tensile strength between its two brands (Brand A and Brand B) of fishing hooks. Tensile strength is typically measured in pascals (Pa). Brand A is considered to be a premiere hook and it costs much more than Brand B which is considered a standard hook. Because the specific alloys used in the making of the hook steel are proprietary, the CEO has obtained the cooperation of the company that supplies the raw materials (rolls of high carbon steel wire) for the hooks, and has obtained the hypothetical data given in Figure 5.19:

PREMIERE (MPa)	STANDARD (MPa)
761	740
760	738
755	736
763	742
765	747
745	744
758	739
761	730
763	737
757	743
750	741
759	738
764	725
767	745
762	743
756	
758	

Fig. 5.19 Worksheet Data for Chapter 5: Practice Problem #3

(a) State the null hypothesis and the research hypothesis on an Excel spreadsheet.
(b) Find the standard error of the difference between the means using Excel
(c) Find the critical t value using Appendix E, and enter it on your spreadsheet.
(d) Perform a t-test on these data using Excel. What is the value of t that you obtain?
(e) State your result on your spreadsheet.
(f) State your conclusion in plain English on your spreadsheet.
(g) Save the file as: FISH13

References

Keller G. Statistics for management and economics. 8th ed. Mason: South-Western Cengage Learning; 2009.
Wheater C, Cook P. Using statistics to understand the environment. New York: Routledge; 2000.

Chapter 6
Correlation and Simple Linear Regression

There are many different types of "correlation coefficients", but the one we will use in this book is the Pearson product-moment correlation which we will call: r.

6.1 What is a "Correlation?"

Basically, a correlation is a number between -1 and $+1$ that summarizes the relationship between two variables, which we will call X and Y.

A correlation can be either positive or negative. *A positive correlation means that as X increases, Y increases. A negative correlation means that as X increases, Y decreases.* In statistics books, this part of the relationship is called the *direction* of the relationship (i.e., it is either positive or negative).

The correlation also tells us the *magnitude* of the relationship between X and Y. As the correlation approaches closer to $+1$, we say that the relationship is *strong and positive.*

As the correlation approaches closer to -1, we say that the relationship is *strong and negative.*

A zero correlation means that there is no relationship between X and Y. This means that neither X nor Y can be used as a predictor of the other.

A good way to understand what a correlation means is to see a "picture" of the scatterplot of points produced in a chart by the data points. Let's suppose that you want to know if variable X can be used to predict variable Y. We will place *the predictor variable X on the x-axis* (the horizontal axis of a chart) and *the criterion variable Y on the y-axis* (the vertical axis of a chart). Suppose, further, that you have collected data given in the scatterplots below (see Fig. 6.1 through Fig. 6.6).

Figure 6.1 shows the scatterplot for a perfect positive correlation of $r = +1.0$. This means that you can perfectly predict each y-value from each x-value because the data points move "upward-and-to-the-right" along a perfectly-fitting straight line (see Fig. 6.1)

T.J. Quirk et al., *Excel 2010 for Physical Sciences Statistics: A Guide to Solving Practical Problems*, DOI 10.1007/978-3-319-00630-7_6,
© Springer International Publishing Switzerland 2013

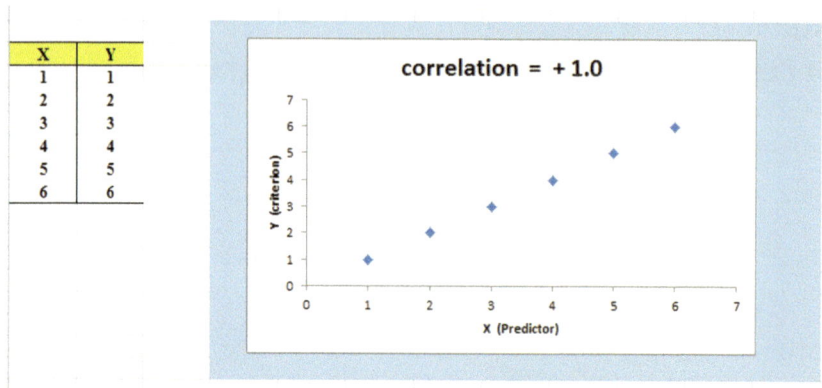

Fig. 6.1 Example of a Scatterplot for a Perfect, Positive Correlation (r = +1.0)

Figure 6.2 shows the scatterplot for a moderately positive correlation of $r = +.54$. This means that each x-value can predict each y-value moderately well because you can draw a picture of a "football" around the outside of the data points that move upward-and-to-the-right, but not along a straight line (see Fig. 6.2).

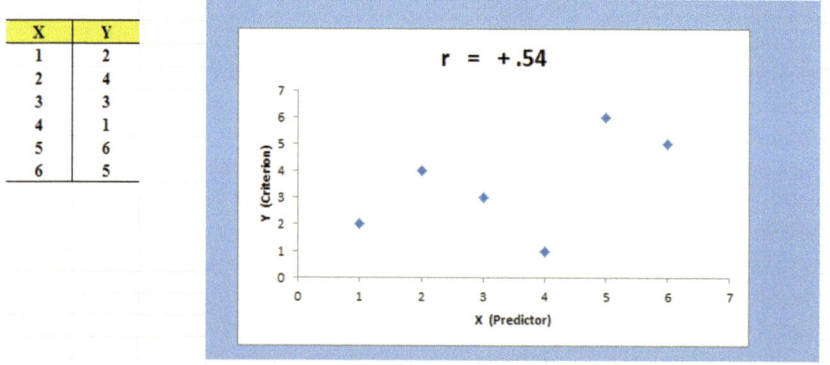

Fig. 6.2 Example of a Scatterplot for a Moderate, Positive Correlation (r = + .54)

Figure 6.3 shows the scatterplot for a low, positive correlation of $r = +.09$. This means that each x-value is a poor predictor of each y-value because the "picture" you could draw around the outside of the data points approaches a circle in shape (see Fig. 6.3)

X	Y
1	2
2	4
3	6
4	1
5	5
6	3

Fig. 6.3 Example of a Scatterplot for a Low, Positive Correlation (r = +.09)

We have not shown a Figure of a zero correlation because it is easy to imagine what it looks like as a scatterplot. A zero correlation of $r = .00$ means that there is no relationship between X and Y and the "picture" drawn around the data points would be a perfect circle in shape, indicating that you cannot use X to predict Y because these two variables are not correlated with one another.

Figure 6.4 shows the scatterplot for a low, negative correlation of $r = -.09$ which means that each X is a poor predictor of Y in an inverse relationship, meaning that as X increases, Y decreases (see Fig. 6.4). In this case, it is a negative correlation because the "football" you could draw around the data points slopes down and to the right.

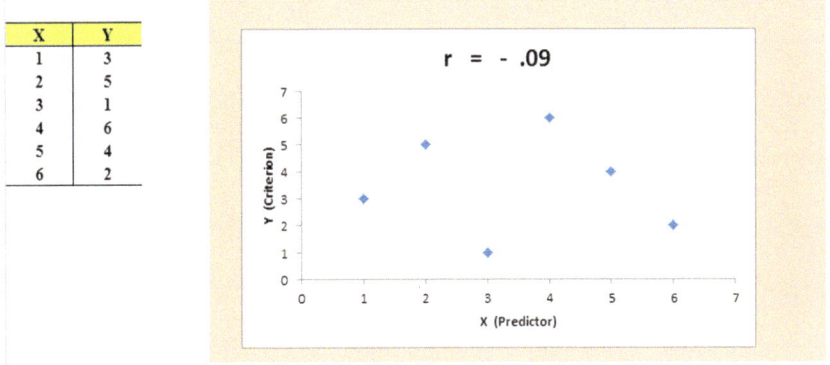

X	Y
1	3
2	5
3	1
4	6
5	4
6	2

Fig. 6.4 Example of a Scatterplot for a Low, Negative Correlation (r = −.09)

Figure 6.5 shows the scatterplot for a moderate, negative correlation of $r = -.54$ which means that X is a moderately good predictor of Y, although there is an inverse relationship between X and Y (i.e., as X increases, Y decreases; see Fig. 6.5). In this case, it is a negative correlation because the "football" you could draw around the data points slopes down and to the right.

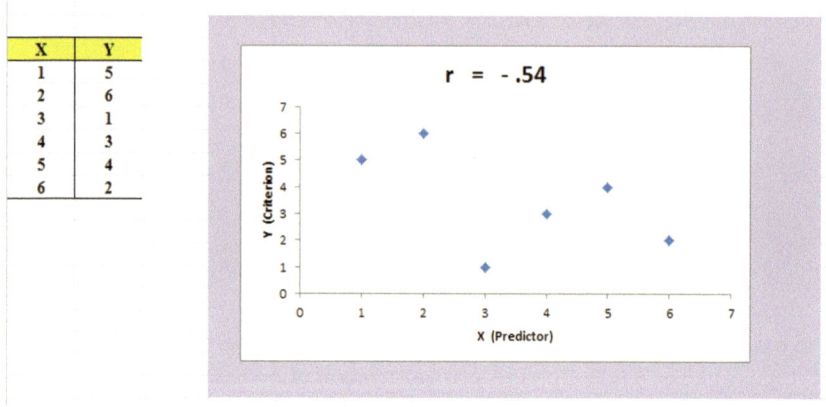

Fig. 6.5 Example of a Scatterplot for a Moderate, Negative Correlation (r = −.54)

Figure 6.6 shows a perfect negative correlation of $r = -1.0$ which means that X is a perfect predictor of Y, although in an inverse relationship such that as X increases, Y decreases. The data points fit perfectly along a downward-sloping straight line (see Fig. 6.6)

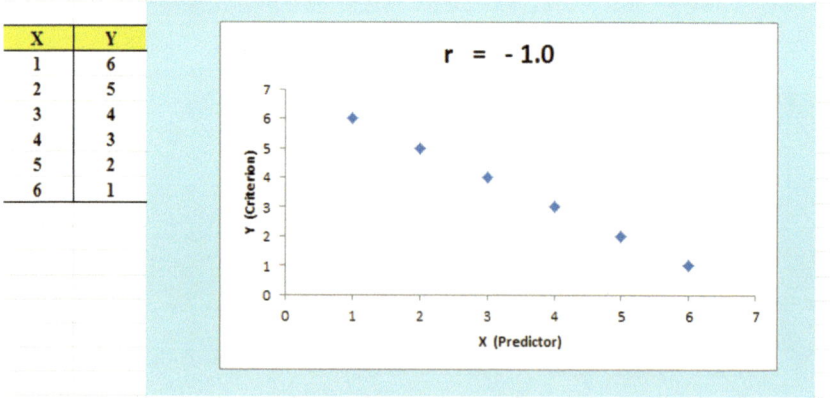

Fig. 6.6 Example of a Scatterplot for a Perfect, Negative Correlation (r = −1.0)

Let's explain the formula for computing the correlation r so that you can understand where the number summarizing the correlation came from.

In order to help you to understand *where* the correlation number that ranges from -1.0 to $+1.0$ comes from, we will walk you through the steps involved to use the formula as if you were using a pocket calculator. This is the one time in this book that we will ask you to use your pocket calculator to find a correlation, but knowing how the correlation is computed step-by-step will give you the opportunity to understand *how* the formula works in practice.

To do that, let's create a situation in which you need to find the correlation between two variables.

Suppose that you wanted to find out if there was a relationship between high school grade-point average (HSGPA) and freshman GPA (FRGPA) for Chemistry majors at a College of Science and Technology. You have decided to call HSGPA the x-variable (i.e., the predictor variable) and FRGPA as the y-variable (i.e., the criterion variable) in your analysis. To test your Excel skills, you take a random sample of freshmen Chemistry majors at the end of their freshman year and record their GPA. The hypothetical data for eight students appear in Fig. 6.7. *(Note: We are using only one decimal place for these GPAs in this example to simplify the mathematical computations).*

	A	B	C	D
1				
2		**X**	**Y**	
3	**Student**	**High School GPA**	**FROSH GPA**	
4	1	2.8	2.9	
5	2	2.5	2.8	
6	3	3.1	2.8	
7	4	3.5	3.2	
8	5	2.4	2.6	
9	6	2.6	2.3	
10	7	2.4	2.1	
11	8	3.6	3.2	
12				
13	**n**	8	8	
14	**MEAN**	2.86	2.74	
15	**STDEV**	0.48	0.39	
16				

Fig. 6.7 Worksheet Data for High School GPA and Frosh GPA (Practical Example)

Notice also that we have used Excel to find the sample size for both variables, X and Y, and the MEAN and STDEV of both variables. (You can practice your Excel skills by seeing if you get this same results when you create an Excel spreadsheet for these data).

Now, let's use the above table to compute the correlation *r* between HSGPA and FRGPA using your pocket calculator.

6.1.1 Understanding the Formula for Computing a Correlation

Objective: To understand the formula for computing the correlation r

The formula for computing the correlation r is as follows:

$$r = \frac{\frac{1}{n-1}\Sigma(X - \bar{X})(Y - \bar{Y})}{S_x S_y} \tag{6.1}$$

This formula looks daunting at first glance, but let's "break it down into its steps" to understand how to compute the correlation r.

6.1.2 Understanding the Nine Steps for Computing a Correlation, r

Objective: To understand the nine steps of computing a correlation r

The nine steps are as follows:

Step	Computation	Result
1	Find the sample size n by noting the number of students	8
2	Divide the number 1 by the sample size minus 1 (i.e., 1 / 7)	0.14286
3	For each student, take the HSGPA and subtract the mean HSGPA for the 8 students and call this $(X - \bar{X})$ (For example, for student # 6, this would be: 2.6 − 2.86) Note: With your calculator, this difference is −0.26, but when Excel uses 16 decimal places for every computation, this result could be slightly different for each student	− 0.26
4	For each student, take the FRGPA and subtract the mean FRGPA for the 8 students and call this $(Y - \bar{Y})$ (For example, for student # 6, this would be: 2.3 − 2.74)	− 0.44
5	Then, for each student, multiply $(X - \bar{X})$ times $(Y - \bar{Y})$ (For example, for student # 6 this would be: $(-0.26) \times (-0.44)$)	+ 0.1144
6	Add the results of $(X - \bar{X})$ times $(Y - \bar{Y})$ for the 8 students	+ 1.09

Steps 1−6 would produce the Excel table given in Fig. 6.8.

	A	B	C	D	E	F	G
1							
2		X	Y				
3	Student	High School GPA	FROSH GPA	$X - \bar{X}$	$Y - \bar{Y}$	$(X - \bar{X})(Y - \bar{Y})$	
4	1	2.8	2.9	-0.06	0.16	-0.01	
5	2	2.5	2.8	-0.36	0.06	-0.02	
6	3	3.1	2.8	0.24	0.06	0.01	
7	4	3.5	3.2	0.64	0.46	0.29	
8	5	2.4	2.6	-0.46	-0.14	0.06	
9	6	2.6	2.3	-0.26	-0.44	0.11	
10	7	2.4	2.1	-0.46	-0.64	0.29	
11	8	3.6	3.2	0.74	0.46	0.34	
12						-------	
13	n	8	8		Total	1.09	
14	MEAN	2.86	2.74				
15	STDEV	0.48	0.39				

Fig. 6.8 Worksheet for Computing the Correlation, r

Notice that when Excel multiplies a minus number by a minus number, the result is a plus number (for example for student #7: $(-0.46)\times(-0.64) = +0.29$). And when Excel multiplies a minus number by a plus number, the result is a negative number (for example for student #1: $(-0.06)x (+0.16) = -0.01$.

Note: Excel computes all computation to 16 decimal places. So, when you check your work with a calculator, you frequently get a slightly different answer than Excel's answer.

For example, when you compute above:

$$(X - \bar{X}) \times (Y - \bar{Y}) \text{ for student #2, your calculator gives:}$$
$$(-0.36) \times (+0.06) = -0.0216$$
(6.2)

As you can see from the table, Excel's answer is −0.02
which is really *more accurate* because Excel uses 16 decimal
places for every number, even though only two decimal places are shown in Figure 6.8.

You should also note that when you do Step 6, you have to be careful to add all of the positive numbers first to get +1.10 and then add all of the negative numbers second to get −0.03 , so that when you subtract these two numbers you get +1.07 as your answer to Step 6. When you do these computations using Excel, this total figure will be +1.09 because Excel carries every number and computation out to 16 decimal places which is much more accurate than your calculator.

Step	Computation	Result
7	Multiply the answer for step 2 above by the answer for step 6 (0.14286 × 1.09)	0.1557
8	Multiply the STDEV of X times the STDEV of Y (0.48 × 0.39)	0.1872
9	Finally, divide the answer from step 7 by the answer from step 8 (0.1557 divided by 0.1872)	0.83

This number of *0.83* is the correlation between HSGPA (X) and FRGPA (Y) for these 8 students. The number +*0.83* means that there is a strong, positive correlation between these two variables. That is, as HSGPA increases, FRGPA increases. For a more detailed discussion of correlation, see Ledholter and Hogg (2010) and McCleery, Watt, and Hart (2007).

You could also use the results of the above table in the formula for computing the correlation r in the following way:

correlation $r = [(1 / (n-1)) \times \sum (X - \overline{X})(Y - \overline{Y})] / (STDEV_x \times STDEV_y)$
correlation $r = [(1/7) \times 1.09] / [(.48) \times (.39)]$
correlation $= r = 0.83$

When you use Excel for these computations, you obtain a slightly different correlation of +0.82 because Excel uses 16 decimal places for all numbers and computations and is, therefore, more accurate than your calculator.

Now, let's discuss how you can use Excel to find the correlation between two variables in a much simpler, and much faster, fashion than using your calculator.

6.2   Using Excel to Compute a Correlation Between Two Variables

Objective: To use Excel to find the correlation between two variables

Suppose that you worked for a car manufacturing company and that you were asked to study the relationship between the weight of 4-door sedans and the fuel consumption they used to drive 150 miles. Suppose, further, that you have obtained 12 sedans, all 2013 models, and have hired drivers to drive 150 miles from Forest Park in St. Louis, Missouri, toward Kansas City, Missouri, on a specified route and at a specified set of speeds. The drivers were all about the same weight.

To test your Excel skills, you have organized the resulting data into a table in which the weight of the cars was measured in thousands of pounds, and the number of gallons of gasoline used in the drive by each car was recorded. The hypothetical data appear in Fig. 6.9.

WEIGHT OF 4-DOOR SEDANS VS. NO. OF GALLONS USED TO DRIVE 150 MILES

Is there a relationship between the weight of a 4-door sedan and the number of gallons used to drive 150 miles?

WEIGHT (thousands of pounds)	NO. OF GALLONS USED
2.1	5.1
2.3	5.3
2.5	5.2
2.6	5.6
2.7	5.1
3.2	6.1
3.2	6.7
3.4	6.8
3.5	6.8
3.6	6.7
3.8	6.5
4.1	6.9

Fig. 6.9 Worksheet Data for Weight and Number of Gallons Used (Practical Example)

Important note: Note that the weight of the cars is recorded in thousands of pounds, so that a car that weighed 3500 pounds would be recorded as 3.5 in this table.

You want to determine if there is a *relationship* between the weight of the cars and their fuel consumption, and you decide to use a correlation to determine this relationship. Let's call the weightof the cars the predictor, X, and the number of gallons used, the criterion, Y.

Create an Excel spreadsheet with the following information:

A3:   WEIGHT OF 4-DOOR SEDANS VS. NO. OF GALLONS USED TO DRIVE 150 MILES
B5:   Is there a relationship between the weight of a 4-door sedan
B6:   and the number of gallons used to drive 150 miles?
B8:   WEIGHT (thousands of pounds)
C8:   NO. OF GALLONS USED
B9:   2.1
C9:   5.1

Next, change the width of Columns B and C so that the information fits inside the cells.

Now, complete the remaining figures in the table given above so that B20 is 4.1 and C20 is 6.9. (Be sure to double-check your figures to make sure that they are correct!) Then, center the information in all of these cells.

A22:   n
A23:   mean
A24:   stdev

Next, define the "name" to the range of data from B9:B20 as: weight
We discussed earlier in this book (see Sect. 1.4.4)how to "name a range of data," but here is a reminder of how to do that:
To give a "name" to a range of data:

Click on the top number in the range of data and drag the mouse
down to the bottom number of the range.

For example, to give the name: "weight" to the cells: B9:B20, click on B9, and drag the pointer down to B20 so that the cells B9:B20 are highlighted on your computer screen. Then, click on:

Formulas
Define name (top center of your screen)
weight(enter this in the Name box; see Fig. 6.10)

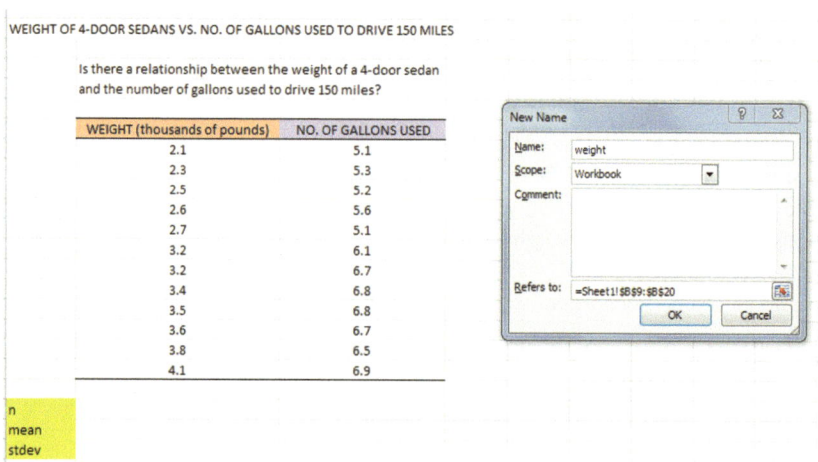

Fig. 6.10   Dialogue Box for Naming a Range of Data as: "weight"

OK

Now, repeat these steps to give the name: gallons to C9:C20
Finally, click on any blank cell on your spreadsheet to "deselect" cells C9:C20 on
   your computer screen.

Now, complete the data for these sample sizes, means, and standard deviations in columns B and C so that B23 is 3.08, and C24 is 0.75 (use two decimals for the means and standard deviations; see Fig. 6.11)

WEIGHT OF 4-DOOR SEDANS VS. NO. OF GALLONS USED TO DRIVE 150 MILES

Is there a relationship between the weight of a 4-door sedan and the number of gallons used to drive 150 miles?

WEIGHT (thousands of pounds)	NO. OF GALLONS USED
2.1	5.1
2.3	5.3
2.5	5.2
2.6	5.6
2.7	5.1
3.2	6.1
3.2	6.7
3.4	6.8
3.5	6.8
3.6	6.7
3.8	6.5
4.1	6.9

n	12	12
mean	3.08	6.07
stdev	0.63	0.75

Fig. 6.11 Example of Using Excel to Find the Sample Size, Mean, and STDEV

Objective: Find the correlation between weight and gallons used

B26: correlation
C26: =correl(weight,gallons); see Fig. 6.12

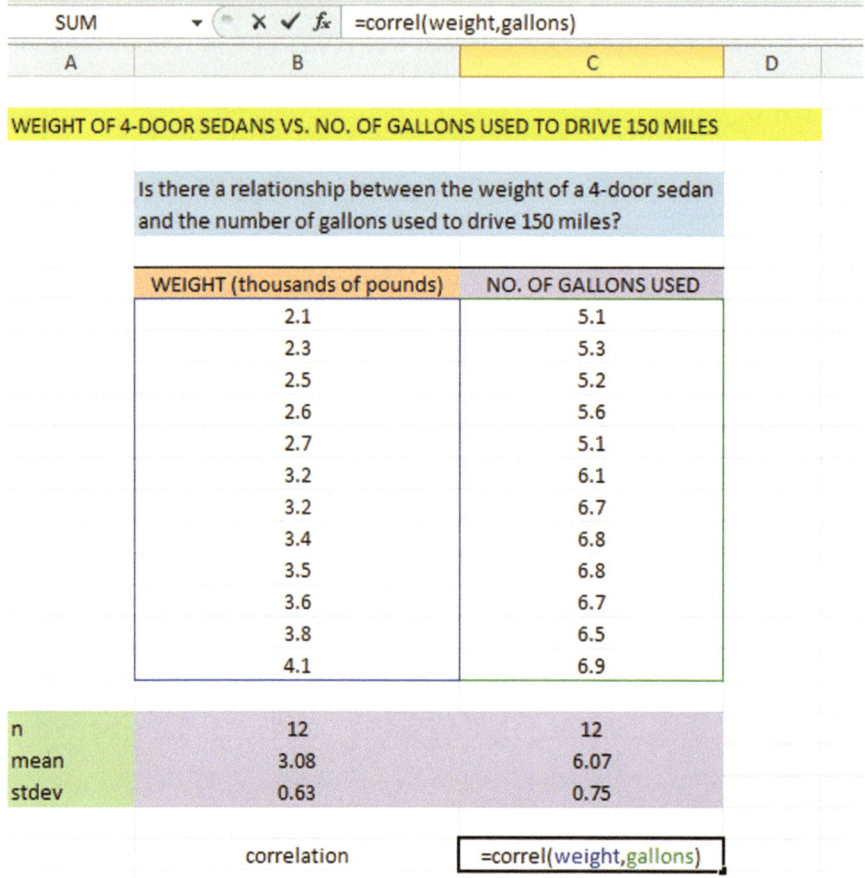

SUM	▾	✕ ✓ ƒx	=correl(weight,gallons)	
A	B		C	D

WEIGHT OF 4-DOOR SEDANS VS. NO. OF GALLONS USED TO DRIVE 150 MILES

Is there a relationship between the weight of a 4-door sedan and the number of gallons used to drive 150 miles?

WEIGHT (thousands of pounds)	NO. OF GALLONS USED
2.1	5.1
2.3	5.3
2.5	5.2
2.6	5.6
2.7	5.1
3.2	6.1
3.2	6.7
3.4	6.8
3.5	6.8
3.6	6.7
3.8	6.5
4.1	6.9

n	12	12
mean	3.08	6.07
stdev	0.63	0.75
	correlation	=correl(weight,gallons)

Fig. 6.12 Example of Using Excel's =correl Function to Compute the Correlation Coefficient

Hit the Enter key to compute the correlation

C26:   format this cell to two decimals

Note that the equal sign in =correl(weight,gallons) in C26 tells Excel that you are going to use a formula in this cell.

The correlation between weight (X) and the number of gallons used (Y) is +.91, a very strong positive correlation. This means that you have evidence that there is a strong relationship between these two variables. In effect, the higher the weight, the more gallons needed to drive 150 miles.

Save this file as: GALLONS3

The final spreadsheet appears in Fig. 6.13.

	A	B	C	D	
2					
3	WEIGHT OF 4-DOOR SEDANS VS. NO. OF GALLONS USED TO DRIVE 150 MILES				
4					
5		Is there a relationship between the weight of a 4-door sedan			
6		and the number of gallons used to drive 150 miles?			
7					
8		WEIGHT (thousands of pounds)	NO. OF GALLONS USED		
9		2.1	5.1		
10		2.3	5.3		
11		2.5	5.2		
12		2.6	5.6		
13	GALLONS3	2.7	5.1		
14		3.2	6.1		
15		3.2	6.7		
16		3.4	6.8		
17		3.5	6.8		
18		3.6	6.7		
19		3.8	6.5		
20		4.1	6.9		
21					
22	n	12	12		
23	mean	3.08	6.07		
24	stdev	0.63	0.75		
25					
26		correlation	0.91		

Fig. 6.13 Final Result of Using the =correl Function to Compute the Correlation Coefficient

6.3   Creating a Chart and Drawing the Regression Line onto the Chart

This section deals with the concept of "linear regression." Technically, the use of a simple linear regression model (i.e., the word "simple" means that only one predictor, X, is used to predict the criterion, Y) requires that the data meet the following four assumptions if that statistical model is to be used:

1. The underlying relationship between the two variables under study (X and Y) is *linear* in the sense that a straight line, and not a curved line, can fit among the data points on the chart.
2. The errors of measurement are independent of each other (e.g. the errors from a specific time period are sometimes correlated with the errors in a previous time period).

3. The errors fit a normal distribution of Y-values at each of the X-values.
4. The variance of the errors is the same for all X-values (i.e., the variability of the Y-values is the same for both low and high values of X).

A detailed explanation of these assumptions is beyond the scope of this book, but the interested reader can find a detailed discussion of these assumptions in Levine *et al.* (2011, pp. 529-530).

Now, let's create a chart summarizing these data.

Important note: Whenever you are preparing a chart, we strongly recommend that you put the predictor variable (X) on the left, and the criterion variable (Y) on the right in your Excel spreadsheet, so that you do not get these variables backwards in your Excel steps and make a mess of the problem in your computations. If you do this as a habit, you will save yourself a lot of grief.

Let's suppose that you would like to use weight of the car as the predictor variable, and that you would like to use it to predict the number of gallons needed to drive 150 miles. Since the correlation between these two variables is +.91, this shows that there is a strong, positive relationship and that weight is a good predictor of the number of gallons needed to drive 150 miles.

1. Open the file that you saved earlier in this chapter:
   GALLONS3

6.3.1   Using Excel to Create a Chart and the Regression Line Through the Data Points

Objective: To create a chart and the regression line summarizing the relationship between weight and gallons used

2. Click and drag the mouse to highlight both columns of numbers (B9:C20), *but do not highlight the labels above the data points.*

   Highlight the data set: B9:C20
   Insert (top left of screen)
   Scatter (at top of screen)
   Click on top left chart icon under "scatter" (see Fig. 6.14)

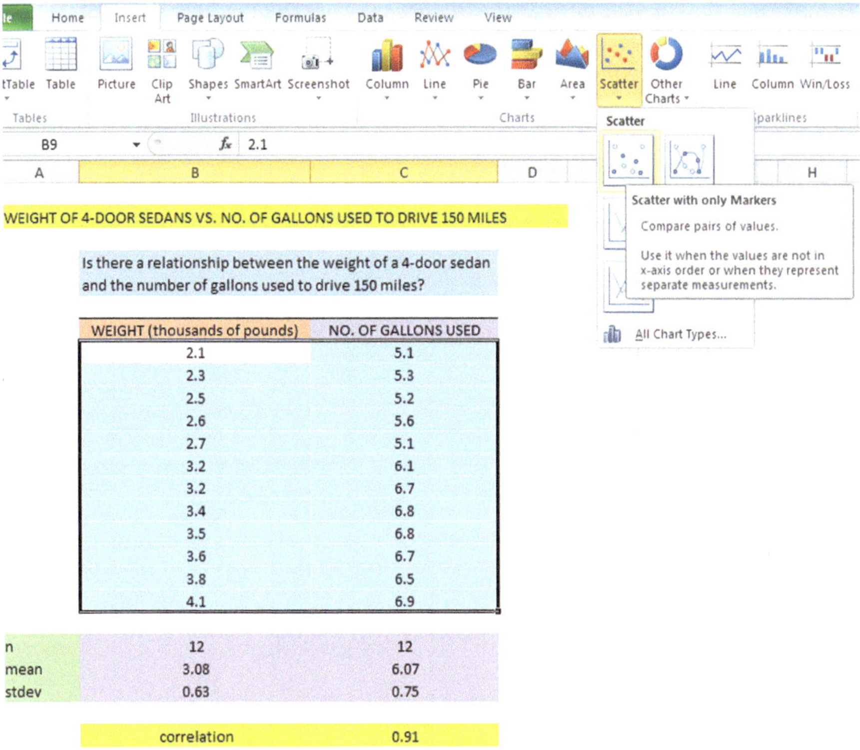

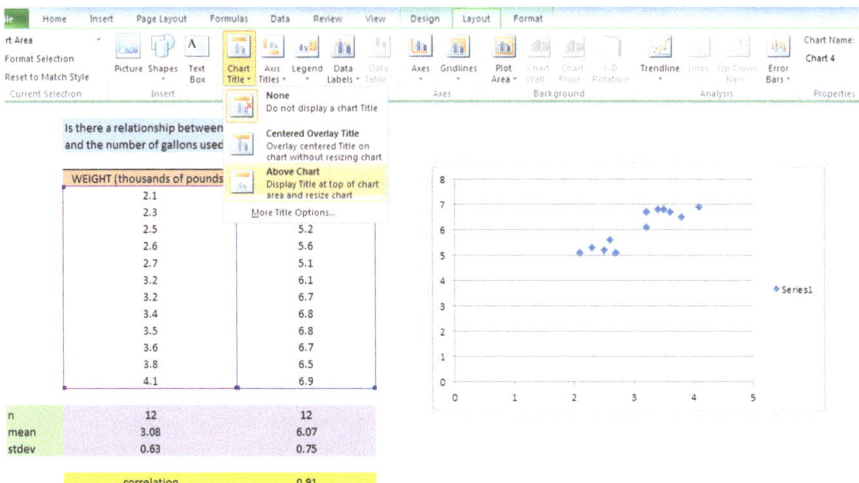

Fig. 6.14 Example of Inserting a Scatter Chart into a Worksheet

Layout (top right of screen under Chart Tools)
Chart title (top of screen)
Above chart (see Fig. 6.15)

Fig. 6.15 Example of Layout / Chart Title / Above Chart Commands

Enter this title in the title box (it will appear to the right of "Chart f_x" at the top of your screen):

RELATIONSHIP BETWEEN WEIGHT AND NO. OF GALLONS USED(see Fig. 6.16)

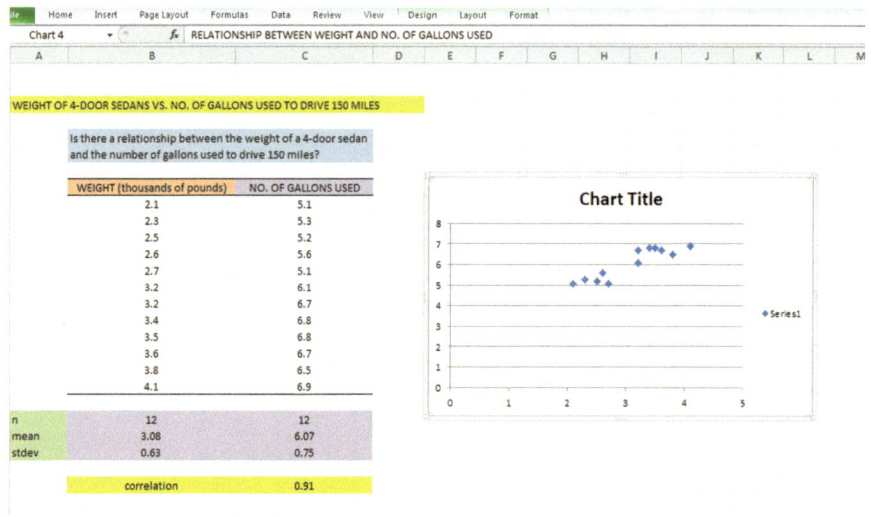

Fig. 6.16   Example of Inserting the Chart title Above the Chart

Hit the enter key to place this title above the chart
Click on *any white space outside of the top title but inside the chart* to "deselect" this chart title

Axis titles (at top of screen)
Primary Horizontal Axis title
Title below axis (see Fig. 6.17)

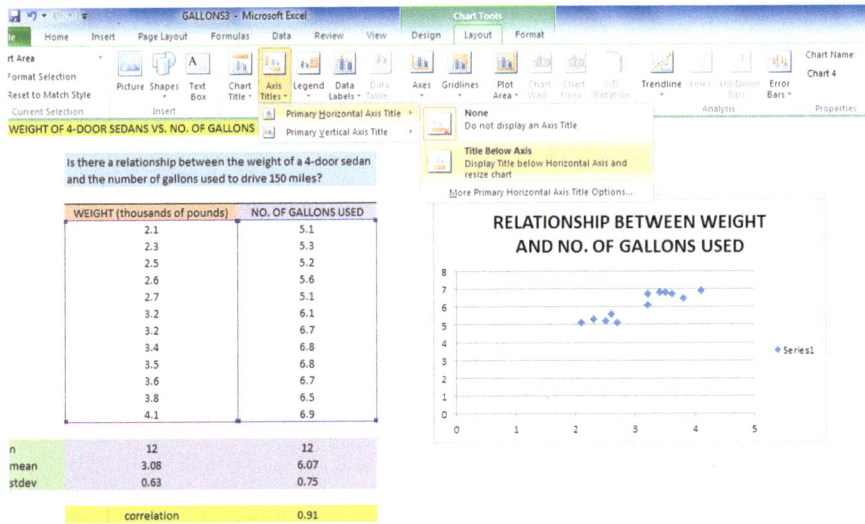

Fig. 6.17   Example of Creating the x-axis Title in a Chart

Now, enter this x-axis title in the "Axis Title Box" at the top of your screen:

WEIGHT (thousands of pounds)

Next, hit the enter key to place this x-axis title at the bottom of the chart

Click on *any white space inside the chart but outside of this x-axis title* to "deselect" the x-axis title

Axis Titles (top center of screen)

Primary Vertical Axis Title

Rotated title

Enter this y-axis title in the Axis Title Box at the top of your screen:

NO. OF GALLONS USED

Next, hit the enter key to place this y-axis title along the y-axis

Then, click on *any white space inside the chart but outside this y-axis title* to "deselect" the y-axis title (see Fig. 6.18)

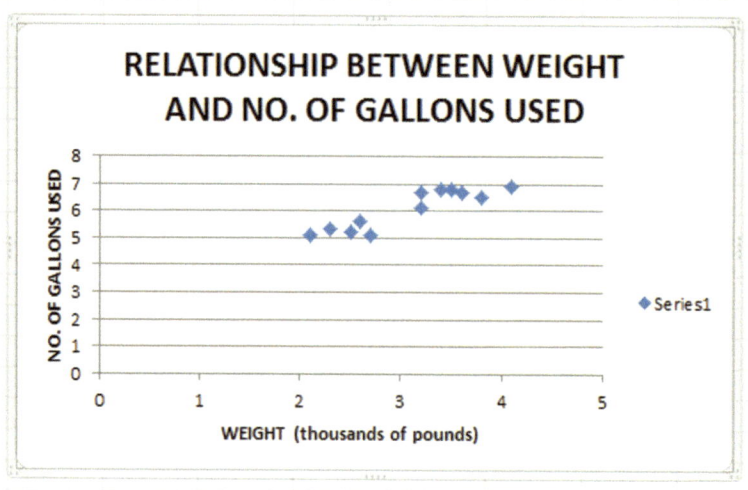

Fig. 6.18 Example of a Chart Title, an x-axis Title, and a y-axis Title

Legend (at top of screen)
None (to turn off the legend "Series 1" at the far right side of the chart)
Gridlines (at top of screen)
Primary Horizontal Gridlines
None (to deselect the horizontal gridlines on the chart)

6.3.1.1   Moving the Chart Below the Table in the Spreadsheet

Objective: To move the chart below the table

Left-click your mouse on *any white space to the right of the top title inside the chart,* keep the left-click down, and drag the chart down and to the left so that the top left corner of the chart is in cell A29, then take your finger off the left-click of the mouse (see Fig. 6.19).

WEIGHT OF 4-DOOR SEDANS VS. NO. OF GALLONS USED TO DRIVE 150 MILES

Is there a relationship between the weight of a 4-door sedan
and the number of gallons used to drive 150 miles?

WEIGHT (thousands of pounds)	NO. OF GALLONS USED
2.1	5.1
2.3	5.3
2.5	5.2
2.6	5.6
2.7	5.1
3.2	6.1
3.2	6.7
3.4	6.8
3.5	6.8
3.6	6.7
3.8	6.5
4.1	6.9

n	12	12
mean	3.08	6.07
stdev	0.63	0.75
	correlation	0.91

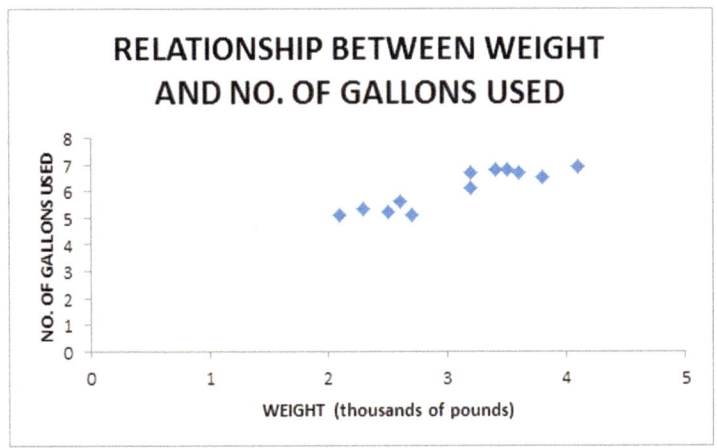

RELATIONSHIP BETWEEN WEIGHT
AND NO. OF GALLONS USED

Fig. 6.19 Example of Moving the Chart Below the Table

6.3.1.2 Making the Chart "Longer" so that it is "Taller"

Objective: To make the chart "longer" so that it is taller

Left-click your mouse on the bottom-center of the chart to create an "up-and-down-arrow" sign, hold the left-click of the mouse down and drag the bottom of the chart down to row 48 to make the chart longer, and then take your finger off the mouse.

6.3.1.3 Making the Chart "Wider"

Objective: Objective: To make the chart "wider"

Put the pointer at the middle of the right-border of the chart to create a "left-to-right arrow" sign, and then left-click your mouse and hold the left-click down while you drag the right border of the chart to the middle of Column H to make the chart wider (see Fig. 6.20).

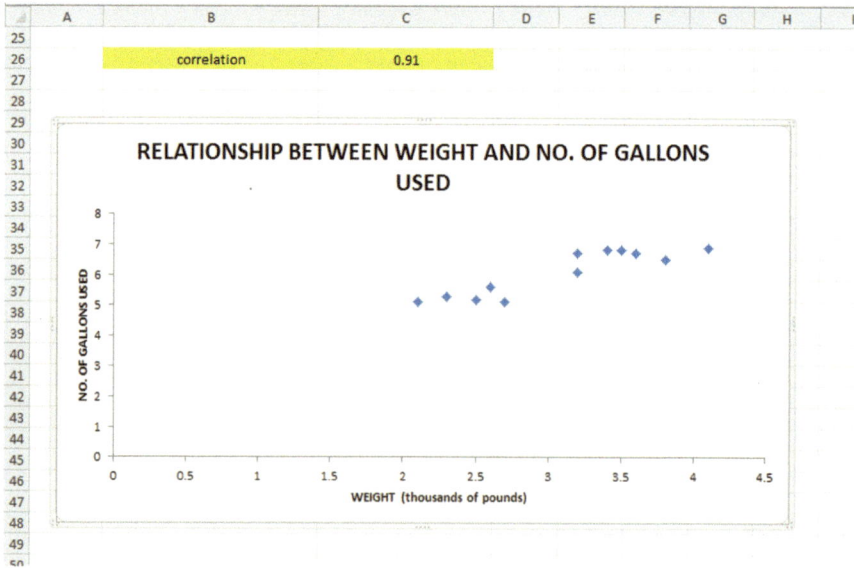

Fig. 6.20 Example of a Chart that is Enlarged to Fit the Cells: A29:H48

Now, let's draw the regression line onto the chart. This regression line is called the "least-squares regression line" and it is the "best-fitting" straight line through the data points.

6.3.1.4   Drawing the Regression Line Through the Data Points in the Chart

> Objective: To draw the regression line through the data points on the chart

Right-click on any one of the data points inside the chart
Add Trendline (see Fig. 6.21)

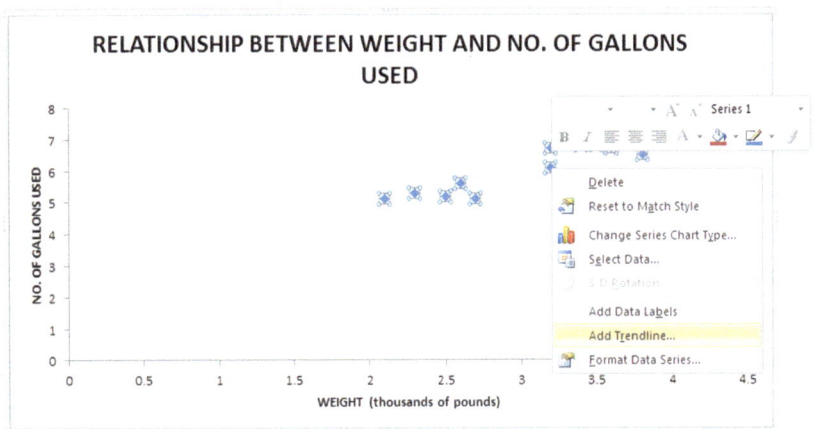

Fig. 6.21   Dialogue Box for Adding a Trendline to the Chart

Linear (be sure the "linear" button on the left is selected; see Fig. 6.22)

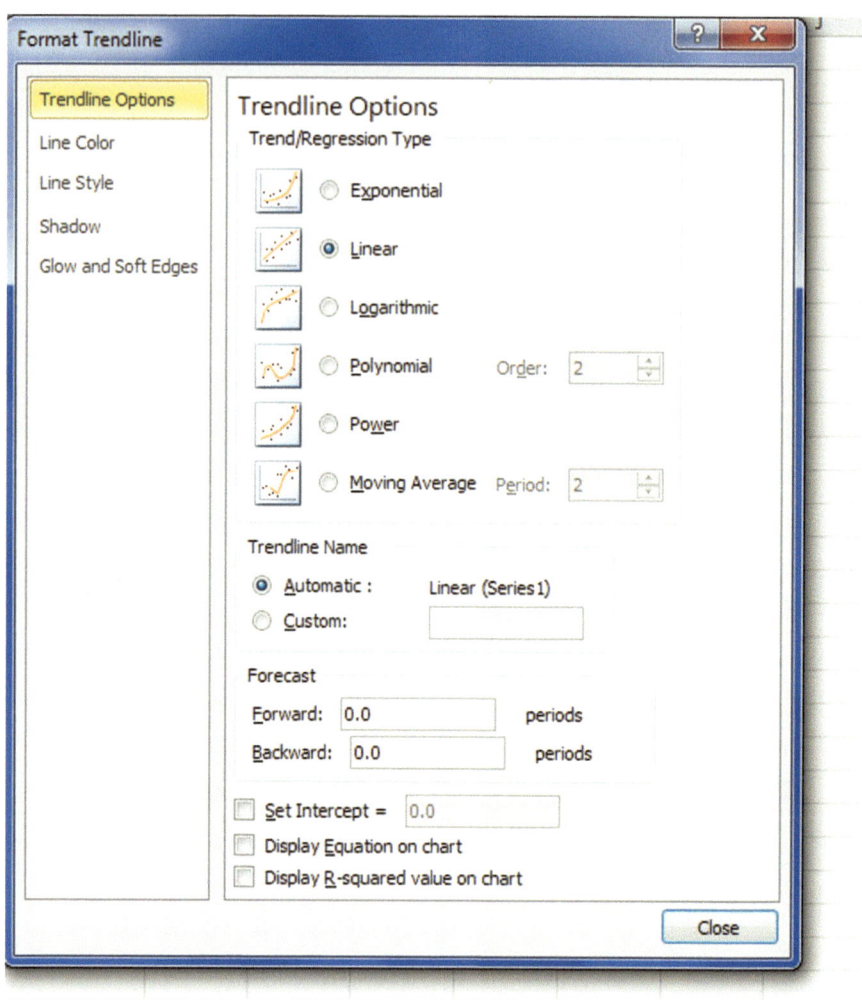

Fig. 6.22 Dialogue Box for a Linear Trendline

Close

Now, click on any blank cell outside the chart to "deselect" the chart

Save this file as: GALLONS4

Note: If you printed this spreadsheet now, it is "too big" to fit onto one page and would "dribble over" onto four pages of printout because the scale needs to be reduced below 100 percent in order for this worksheet to fit onto only one page. You need to complete these next steps below to print out some, or all, of this spreadsheet.

6.4   Printing a Spreadsheet So That the Table and Chart Fit Onto One Page

Objective: To print the spreadsheet so that the table and the chart fit onto one
          page

Page Layout (top of screen)
Change the scale at the middle icon near the top of the screen "Scale to Fit" by
clicking on the down-arrow until it reads "80 %" so that the table and the chart will
fit onto one page on your printout (see Fig. 6.23):

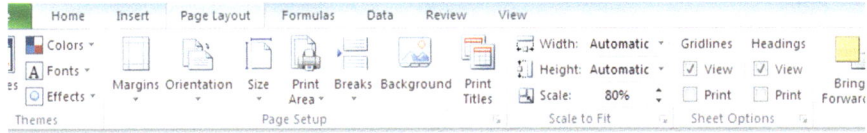

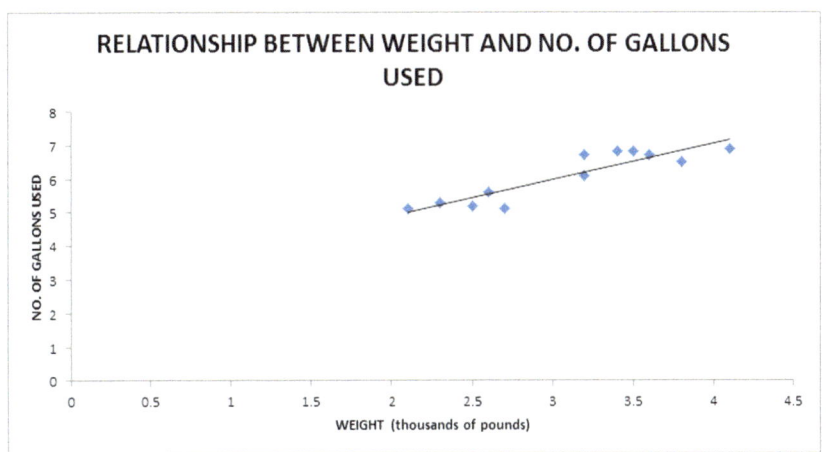

Fig. 6.23 Example of the Page Layout for Reducing the Scale of the Chart to 80 % of Normal Size

File
Print
Print (see Fig. 6.24)

WEIGHT OF 4-DOOR SEDANS VS. NO. OF GALLONS USED TO DRIVE 150 MILES

Is there a relationship between the weight of a 4-door sedan
and the number of gallons used to drive 150 miles?

WEIGHT (thousands of pounds)	NO. OF GALLONS USED
2.1	5.1
2.3	5.3
2.5	5.2
2.6	5.6
2.7	5.1
3.2	6.1
3.2	6.7
3.4	6.8
3.5	6.8
3.6	6.7
3.8	6.5
4.1	6.9

n	12	12
mean	3.08	6.07
stdev	0.63	0.75
	correlation	0.91

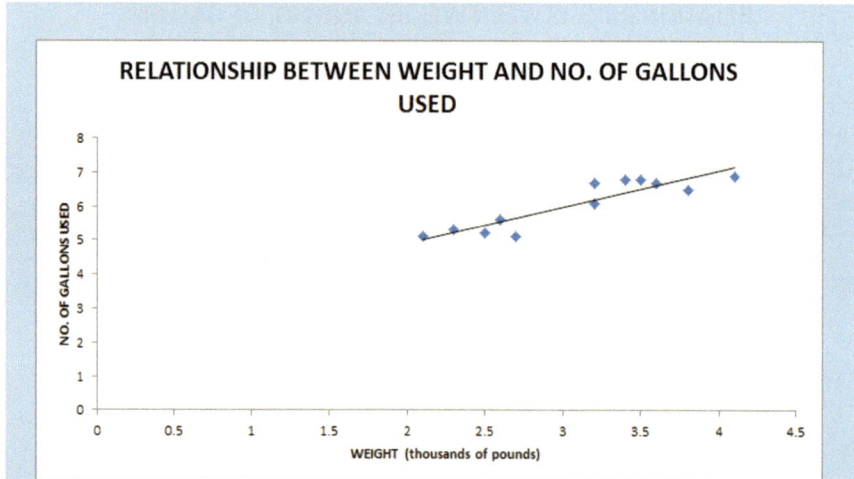

Fig. 6.24 Final Spreadsheet of Regression Line on a Chart (80 % Scale to Fit Size)

Re-save your file as: GALLONS4

6.5 Finding the Regression Equation

The main reason for charting the relationship between X and Y (i.e., weight as X and the number of gallons used as Y in our example) is to see if there is a strong enough relationship between X and Y so that the regression equation that summarizes this relationship can be used to predict Y for a given value of X.

Since we know that the correlation between the weight of the cars and the number of gallons used is $+.91$, this tells us that it makes sense to use weight to predict the number of gallons used based on past data.

We now need to find that regression equation that is the equation of the "best-fitting straight line" through the data points.

> Objective: To find the regression equation summarizing the relationship between X and Y.

In order to find this equation, we need to check to see if your version of Excel contains the "Data Analysis ToolPak" necessary to run a regression analysis.

6.5.1 Installing the Data Analysis ToolPak into Excel

> Objective: To install the Data Analysis ToolPak into Excel

Since there are currently three versions of Excel in the marketplace (2003, 2007, 2010), we will give a brief explanation of how to install the Data Analysis ToolPak into each of these versions of Excel.

6.5.1.1 Installing the Data Analysis ToolPak into Excel 2010

Open a new Excel spreadsheet
Click on: Data (at the top of your screen)

Look at the top of your monitor screen. Do you see the words: "Data Analysis" at the far right of the screen? If you do, the Data Analysis ToolPak for Excel 2010 was correctly installed when you installed Office 2010, and you should skip ahead to Sect. 6.5.2.

If the words: "Data Analysis" are not at the top right of your monitor screen, then the ToolPak component of Excel 2010 was not installed when you installed Office 2010 onto your computer. If this happens, you need to follow these steps:

File
Options
Excel options (creates a dialog box)

Add-Ins
Manage: Excel Add-Ins (at the bottom of the dialog box)
Go
Highlight: Analysis ToolPak (in the Add-Ins dialog box)
OK
Data
(You now should have the words: "Data Analysis" at the top right of your screen)
　　If you get a prompt asking you for the "installation CD," put this CD in the CD drive and click on: OK

Note: If these steps do not work, you should try these steps instead: File / Options (bottom left) / Add-ins / Analysis ToolPak / GO / click to the left of Analysis ToolPak to add a check mark / OK

If you need help doing this, ask your favorite "computer techie" for help.

　　You are now ready to skip ahead to Sect. 6.5.2.

6.5.1.2　Installing the Data Analysis ToolPak into Excel 2007

Open a new Excel spreadsheet
Click on:　Data (at the top of your screen

If the words "Data Analysis" do not appear at the top right of your screen, you
　need to install the Data Analysis ToolPak using the following steps:

Microsoft Office button (top left of your screen)
Excel options (bottom of dialog box)
Add-ins (far left of dialog box)
Go (to create a dialog box for Add-Ins)
Highlight: Analysis ToolPak
OK (If Excel asks you for permission to proceed, click on: Yes)
Data
(You should now have the words: "Data Analysis" at the top right of your screen)

If you need help doing this, ask your favorite "computer techie" for help.

　　You are now ready to skip ahead to Sect. 6.5.2.

6.5.1.3　Installing the Data Analysis ToolPak into Excel 2003

Open a new Excel spreadsheet
Click on: Tools (at the top of your screen)

　　If the bottom of this Tools box says "Data Analysis," the ToolPak has already been installed in your version of Excel and you are ready to find the regression equation. If the bottom of the Tools box does not say "Data Analysis," you need to install the ToolPak as follows:

Click on: File

Options (bottom left of screen)
Add-ins
Analysis Tool Pak (it is directly underneath Inactive
        Application Add-ins near the top of the box)
Go
Click to add a check-mark to the left of analysis Toolpak
OK

*Note: If these steps do not work, try these steps instead: Tools / Add-ins / Click to
the left of analysis ToolPak to add a check mark to the left / OK*

You are now ready to skip ahead to Sect. 6.5.2.

6.5.2 *Using Excel to Find the SUMMARY OUTPUT of Regression*

You have now installed *ToolPak*, and you are ready to find the regression equation for
the "best-fitting straight line" through the data points by using the following steps:

Open the Excel file: *GALLONS4* (if it is not already open on your screen)

Note: *If this file is already open, and there is a gray border around the chart you
need to click on any empty cell outside of the chart to deselect the chart.*

Now that you have installed *Toolpak*, you are ready to find the regression equation
summarizing the relationship between weight and gallons used in your data set.
Remember that you gave the name: *weight* to the X data (the predictor), and the
name: *gallons* to the Y data (the criterion) in a previous section of this chapter (see
Sect. 6.2)

Data (top of screen)
Data analysis (far right at top of screen; see Fig. 6.25)

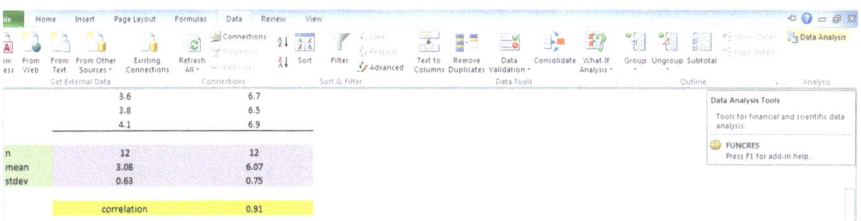

Fig. 6.25 Example of Using the Data / Data Analysis Function of Excel

Scroll down the dialog box using the down arrow and click on: Regression (see
Fig. 6.26)

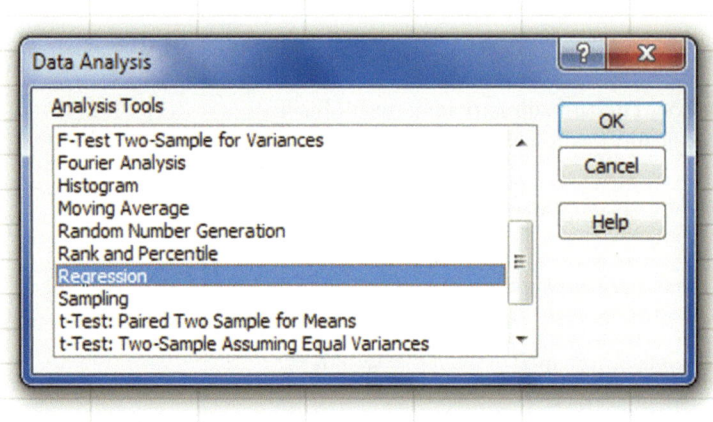

Fig. 6.26 Dialogue Box for Creating the Regression Function in Excel

OK

Input Y Range: gallons
Input X Range: weight

Click on the "button" to the left of Output Range to select this, and enter A50 in the
    box as the place on your spreadsheet to insert the Regression analysis in cell A50
OK
The *SUMMARY OUTPUT* should now be in cells: A50 : I67

Now, make the columns in the Regression Summary Output section of your
spreadsheet *wider* so that you can read all of the column headings clearly.

Now, change the data in the following two cells to Number format (2 decimal
places):

B53
B66

Next, change this cell to four decimal places: B67
Now, change the format for all other numbers that are in decimal format to
number format, three decimal places, and center all numbers within their cells.

Save the resulting file as: GALLONS5

Print the file so that it fits onto one page. (*Hint: Change the scale under "Page
Layout" to 60 % to make it fit).* Your file should be like the file in Fig. 6.27.

WEIGHT OF 4-DOOR SEDANS VS. NO. OF GALLONS USED TO DRIVE 150 MILES

	Is there a relationship between the weight of a 4-door sedan and the number of gallons used to drive 150 miles?	
	WEIGHT (thousands of pounds)	NO. OF GALLONS USED
	2.1	5.1
	2.3	5.3
	2.5	5.2
	2.6	5.6
GALLONS5	2.7	5.1
	3.2	6.1
	3.2	6.7
	3.4	6.8
	3.5	6.8
	3.6	6.7
	3.8	6.5
	4.1	6.9
n	12	12
mean	3.08	6.07
stdev	0.63	0.75
	correlation	0.91

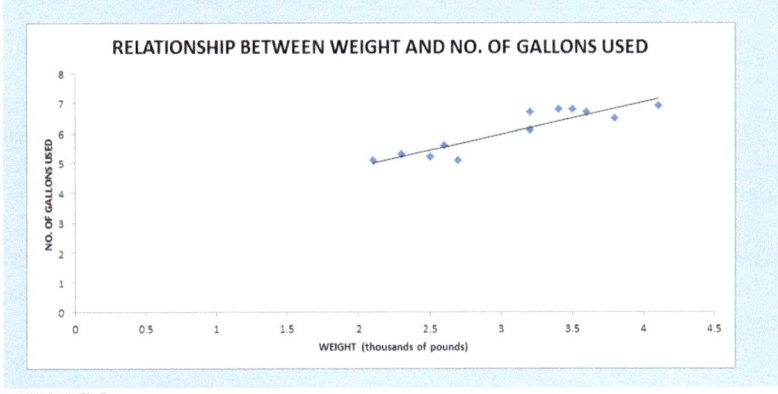

SUMMARY OUTPUT

Regression Statistics	
Multiple R	0.91
R Square	0.827
Adjusted R Square	0.810
Standard Error	0.327
Observations	12

ANOVA

	df	SS	MS	F	Significance F
Regression	1	5.116	5.116	47.765	4.13743E-05
Residual	10	1.071	0.107		
Total	11	6.187			

	Coefficients	Standard Error	t Stat	P-value	Lower 95%	Upper 95%	Lower 95.0%	Upper 95.0%
Intercept	2.75	0.489	5.616	0.000	1.658	3.839	1.658	3.839
X Variable 1	1.0762	0.156	6.911	0.000	0.729	1.423	0.729	1.423

Fig. 6.27 Final Spreadsheet of Correlation and Simple Linear Regression including the SUMMARY OUTPUT for the Data

Note the following problem with the summary output.

Whoever wrote the computer program for this version of Excel made a mistake and gave the name: "Multiple R" to cell A53. This is not correct. Instead, cell A53 should say: "correlation r" since this is the notation that we are using for the correlation between X and Y.

You can now use your printout of the regression analysis to find the regression equation that is the best-fitting straight line through the data points.

But first, let's review some basic terms.

6.5.2.1   Finding the y-intercept, a, of the Regression Line

The point on the y-axis that the regression line would intersect the y-axis if it were extended to reach the y-axis is called the "y-intercept" and *we will use the letter "a" to stand for the y-intercept of the regression line.* The y-intercept on the SUMMARY OUTPUT of Fig. 6.27 is *2.75 and appears in cell B66.* This means that if you were to draw an imaginary line continuing down the regression line toward the y-axis that this imaginary line would cross the y-axis at 2.75. This is why it is called the "y-intercept."

6.5.2.2   Finding the Slope, b, of the Regression Line

The "tilt" of the regression line is called the "slope" of the regression line. It summarizes to what degree the regression line is either above or below a horizontal line through the data points. If the correlation between X and Y were zero, the regression line would be exactly horizontal to the X-axis and would have a zero slope.

If the correlation between X and Y is positive, the regression line would "slope upward to the right" above the X-axis. Since the regression line in Fig. 6.27 slopes upward to the right, the slope of the regression line is +1.0762 as given in cell *B67*. *We will use the notation "b" to stand for the slope of the regression line.* (Note that Excel calls the slope of the line: "X Variable 1" in the Excel printout.)

Since the correlation between weight and gallons used was +.91, you can see that the regression line for these data "slopes upward to the right" through the data. Note that the SUMMARY OUTPUT of the regression line in Fig. 6.27 gives a correlation, r , of +.91 in cell *B53*.

If the correlation between X and Y were negative, the regression line would "slope down to the right" above the X-axis. This would happen whenever the correlation between X and Y is a negative correlation that is between zero and minus one (0 and −1).

6.5.3   Finding the Equation for the Regression Line

To find the regression equation for the straight line that can be used to predict the number of gallons used from the car's weight, we only need two numbers in the SUMMARY OUTPUT in Fig. 6.27: *B66 and B67*.

The format for the regression line is:

$$Y = a + bX \qquad\qquad (6.3)$$

where a = *the y-intercept* (2.75 in our example in cell B66) and b = *the slope of the line* (+1.0762 in our example in cell B67)

Therefore, the equation for the best-fitting regression line for our example is:

$$Y = a + bX$$

$$Y = 2.75 + 1.0762X$$

Remember that Y is the number of gallons used that we are trying to predict, using the weight of the car as the predictor, X.

Let's try an example using this formula to predict the number of gallons used for a car.

6.5.4 Using the Regression Line to Predict the y-value for a Given x-value

Objective: To find the number of gallons predicted for a car that weighed 3,000 pounds (Note: 3,000 pounds, when measured in thousands of pounds, is recorded as 3.0)

Important note: Remember that the weight of the car in thousands of pounds.

Since the weight is 3000 pounds (i.e., X = 3.0 in thousands of pounds), substituting this number into our regression equation gives:

$$Y = 2.75 + 1.0762 \ (3.0)$$
$$Y = 2.75 + 3.23$$
$$Y = 5.98 \text{ gallons of gas needed to drive 150 miles}$$

Important note: If you look at your chart, if you go directly upwards for a weight of 3.0 until you hit the regression line, you see that you hit this line just below 6 on the y-axis to the left when you draw a line horizontal to the x-axis (actually, it is 5.98), the result above for predicting the number of gallons needed for a car weighing 3000 pounds.

Now, let's do a second example and predict the number of gallons needed if we used a weight of 3500 pounds. (Remember: 3500 pounds becomes 3.5 when it is measured in thousands of pounds).

$$Y = 2.75 + 1.0762 \ X$$
$$Y = 2.75 + 1.0762 \ (3.5)$$

$$Y = 2.75 + 3.77$$
$$Y = 6.52 \text{ gallons of gas needed to drive 150 miles}$$

Important note: If you look at your chart, if you go directly upwards for a weight of 3.5 until you hit the regression line, you see that you hit this line between 6 and 7 on the y-axis to the left (actually it is 6.52), the result above for predicting the number of gallons of gas needed for a car that weighed 3500 pounds to drive 150 miles.

For a more detailed discussion of regression, see Black (2010) and McKillup and Dyar (2010).

6.6 Adding the Regression Equation to the Chart

Objective: To Add the Regression Equation to the Chart

If you want to include the regression equation within the chart next to the regression line, you can do that, but a word of caution first.

Throughout this book, we are using the regression equation for one predictor and one criterion to be the following:

$$Y = a + bX \tag{6.3}$$

where a = y-intercept and b = slope of the line

See, for example, the regression equation in Sect. 6.5.3 where the y-intercept was *a* =2.75 and the slope of the line was *b* = + *1.0762* to generate the following regression equation:

$$Y = 2.75 + 1.0762X$$

However, Excel 2010 uses a slightly different regression equation (which is logically identical to the one used in this book) when you add a regression equation to a chart:

$$Y = bX + a \tag{6.4}$$

where a = y-intercept and b = slope of the line

Note that this equation is identical to the one we are using in this book with the terms arranged in a different sequence.

For the example we used in Sect. 6.5.3, Excel 2010 would write the regression equation on the chart as:

$$Y = 1.0762X + 2.75$$

This is the format that will result when you add the regression equation to the chart using Excel 2010 using the following steps:

Open the file: GALLONS5*(that you saved in Sect. 6.5.2)*

Click just *inside* the outer border of the chart in the top right corner to add the "gray border" around the chart in order to "select the chart" for changes you are about to make

Right-click on any of the data-points in the chart

Highlight: Add Trendline

The "Linear button" near the top of the dialog box will be selected (on its left)

Click on: Display Equation on chart (near the bottom of the dialog box; see Fig. 6.28)

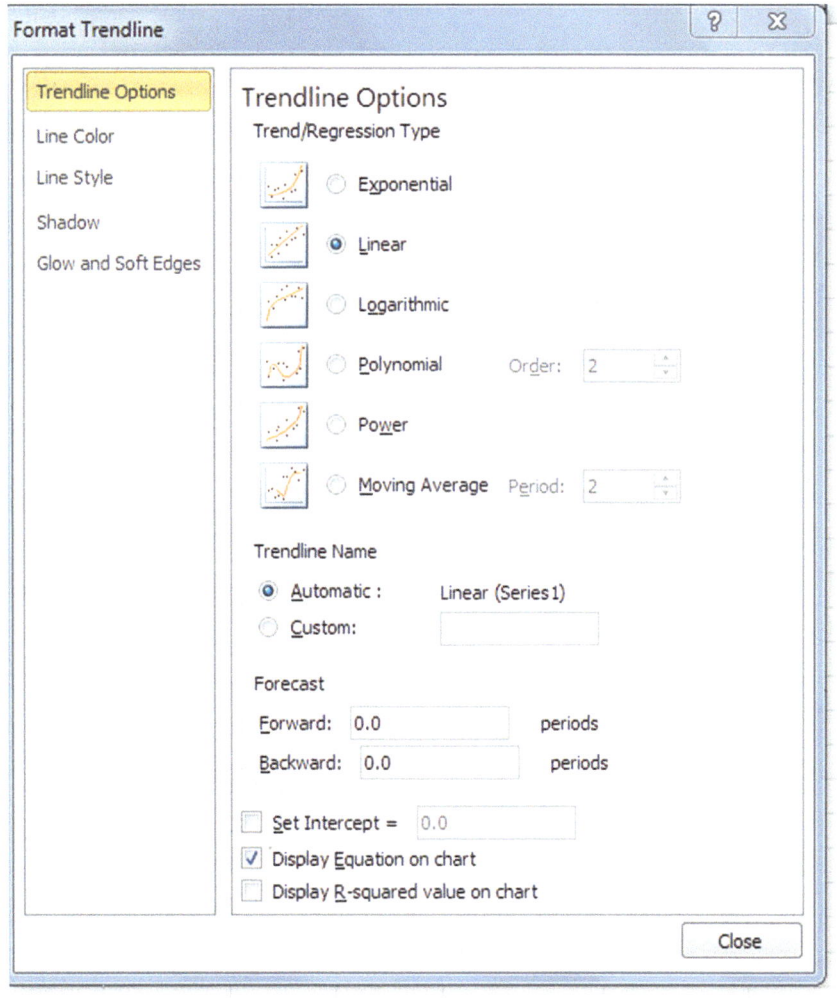

Fig. 6.28 Dialogue Box for Adding the Regression Equation to the Chart Next to the Regression Line on the Chart

Close

Note that the regression equation on the chart is in the following form next to the regression line on the chart (see Fig. 6.29).

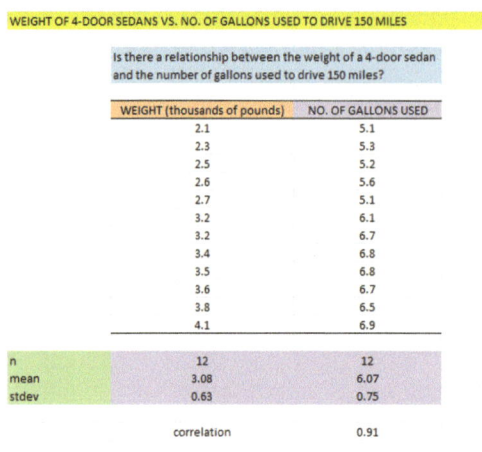

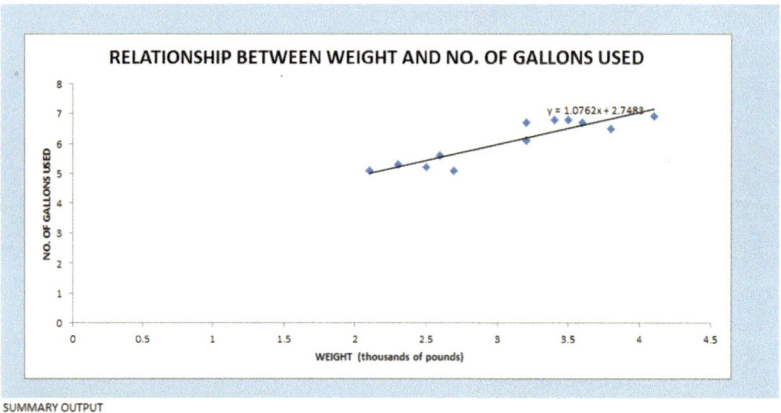

Fig. 6.29 Example of a Chart with the Regression Equation Displayed Next to the Regression Line

$$Y = 1.0762 \ X + 2.75$$

Now, save this file as: GALLONS6

6.7   How to Recognize Negative Correlations in the SUMMARY OUTPUT Table

Important note: Since Excel does not recognize negative correlations in the SUMMARY OUTPUT results, buttreats all correlations as if they were positive correlations (this was a mistake made by the programmer), you need to be careful to note that there may be a negative correlation between X and Y even if the printout says that the correlation is a positive correlation.

You will know that the correlation between X and Y is a negative correlation when these two things occur:

(1) *THE SLOPE, b, IS A NEGATIVE NUMBER. This can only occur when there is a negative correlation.*
(2) *THE CHART CLEARLY SHOWS A DOWNWARD SLOPE IN THE REGRESSION LINE, which can only occur whenthe correlation between X and Y is negative.*

6.8   Printing Only Part of a Spreadsheet Instead of the Entire Spreadsheet

> Objective: To print part of a spreadsheet separately instead of printing the entire spreadsheet

There will be many occasions when your spreadsheet is so large in the number of cells used for your data and charts that you only want to print part of the spreadsheet separately so that the print will not be so small that you cannot read it easily.

We will now explain how to print only part of a spreadsheet onto a separate page by using three examples of how to do that using the file, GALLONS6, that you created in Sect. 6.6: (1) printing only the table and the chart on a separate page, (2) printing only the chart on a separate page, and (3) printing only the SUMMARY OUTPUT of the regression analysis on a separate page.

Note: If the file: GALLONS6 is not open on your screen, you need to open it now.

Let's describe how to do these three goals with three separate objectives:

6.8.1   Printing Only the Table and the Chart on a Separate Page

> Objective: To print only the table and the chart on a separate page

1. Left-click your mouse starting at the top left of the table *in cell A3* and drag the mouse *down and to the right so that all of the table and all of the chart are highlighted in light blue on your computer screen from cell A3 to cell H48* (the light blue cells are called the "selection" cells).
2. File
   Print
   Print Active Sheet (hit the down arrow on the right)
   Print selection
   Print

The resulting printout should contain only the table of the data and the chart resulting from the data.
Then, click on any empty cell in your spreadsheet to deselect the table and chart.

6.8.2 Printing Only the Chart on a Separate Page

Objective: To print only the chart on a separate page
---

1. Click on any "white space" *just inside the outside border of the chart in the top right corner of the chart* to create the gray border around all of the borders of the chart in order to "select" the chart.
2. File
   Print
   Print selected chart
   Print selected chart (again)
   Print

The resulting printout should contain only the chart resulting from the data.

Important note: After each time you print a chart by itself on a separate page, you should immediately click on any white space OUTSIDE the chart to remove the gray border from the border of the chart. When the gray border is on the borders of the chart, this tells Excel that you want to print only the chart by itself. You should do this now!

6.8.3 Printing Only the SUMMARY OUTPUT of the Regression Analysis on a Separate Page

---
Objective: To print only the SUMMARY OUTPUT of the regression analysis on a separate page
---

1. Left-click your mouse at the cell just above SUMMARY OUTPUT in *cell A50* on
   the left of your spreadsheet and drag the mouse *down and to the right* until all of
   the regression output is highlighted in dark blue on your screen from A50 to I67.
2. File
   Print
   Print active sheets (hit the down arrow on the right)
   Print selection
   Print

   The resulting printout should contain only the summary output of the regression
      analysis on a separate page.
   Finally, click on any empty cell on the spreadsheet to "deselect" the regression
      table.

6.9   End-Of-Chapter Practice Problems

1. Suppose that you wanted to study the relationship between the amount of tin
   extracted from a product based on its refluxing time when the product is boiled
   with hydrochloric acid. Is there a relationship between these two variables?
      You have decided to use the refluxing time in minutes(min) as the predictor,
   X, and the amount of tin extracted in milligrams per kilogram(mg/kg) as the
   criterion, Y. To test your Excel skills, you have randomly chosen 12 samples of
   this process, and have recorded the hypothetical scores on these variables in
   Fig. 6.30:

RELATIONSHIP BETWEEN REFLUXING TIME AND AMOUNT OF TIN EXTRACTED

Refluxing Time (min)	Tin Extracted (mg/kg)
35	53
40	56
45	54
50	59
55	56
60	58
65	56
70	54
75	56
80	57
85	59
90	58

Fig. 6.30 Worksheet Data for Chapter 6: Practice Problem #1

Create an Excel spreadsheet and enter the data *using REFLUXING TIME (min) as the independent variable (predictor) and TIN EXTRACTED (mg/kg) as the dependent variable (criterion).*

Important note: When you are trying to find a correlation between two variables, it is important that you place the predictor, X, ON THE LEFT COLUMN in your Excel spreadsheet, and the criterion, Y, IMMEDIATELY TO THE RIGHT OF THE X COLUMN. You should do this every time that you want to use Excel to find a correlation between two variables to check your thinking so that you do not confuse these two variables with one another.

(a) Create the table using Excel, and then use Excel's =correl function to find the correlation between these two variables, and round off the result to two decimal places. Label the correlation and place it beneath the table.

(b) Create an *XY scatterplot* of these two sets of data such that:

- Top title: RELATIONSHIP BETWEEN REFLUXING TIME AND AMOUNT OF TIN EXTRACTED
- x-axis title: REFLUXING TIME (min)
- y-axis title: TIN EXTRACTED (mg/kg)
- re-size the chart so that it is 8 columns wide and 25 rows long
- delete the legend
- delete the gridlines
- move the chart below the table

(c) Create the *least-squares regression line* for these data on the scatterplot and add the regression equation to the chart.

(d) Use Excel to run the regression statistics to find the *equation for the least-squares regression line* for these data and display the results below the chart on your spreadsheet. Use number format (2 decimal places) for the correlation and three decimal places for all the other decimal numbers, including the coefficients.

(e) Print just the input data and the chart so that this information fits onto one page. Then, print the regression output table on a separate page so that it fits onto that separate page.

(f) save the file as: TIN3

Now, answer these questions using your Excel printout:

(1) What is the correlation coefficient *r*?
(2) What is the y-intercept?
(3) What is the slope of the line?
(4) What is the regression equation for these data (use three decimal places for the y-intercept and the slope)?
(5) Use the regression equation to predict the amount of tin extracted you would expect for a refluxing time of 60 minutes.

2. Permafrost is soil, sediment, or rock that is frozen based on its temperature. The ground must remain at or below zero degrees centigrade for two years or more to be called permafrost. It is found at high altitudes, including the Rocky Mountains in the state of Colorado. Permafrost is measured by down-hole depth created by a drill hole in a formation that is used as part of geophysical studies. Suppose that you wanted to study the relationship between down-hole depth (X) and temperature. Suppose that down-hole depth was measured in meters (m) while temperature was measured in degrees centigrade (°C).

Create an Excel spreadsheet and enter the data using DEPTH as the independent (predictor) variable, and TEMPERATURE as the dependent (criterion) variable. You decide to test your Excel skills on a small sample of drill holes using the hypothetical data presented in Fig. 6.31.

DOWN-HOLE DEPTH (meters) VS. TEMPERATURE (degrees centigrade)

DEPTH (m)	TEMPERATURE (°C)
0.1	-3.6
0.3	-3.5
0.6	-2.7
0.9	-2.5
1.4	-2.6
2.2	-2.7
3.2	-2.4
4.8	-0.2
6.8	0.0

Fig. 6.31 Worksheet Data for Chapter 6: Practice Problem #2

Create an Excel spreadsheet and enter the data using DEPTH (meters) as the independent variable (predictor) and TEMPERATURE (degrees centigrade) as the dependent variable (criterion).

(a) create an *XY scatterplot* of these two sets of data such that:

- top title: RELATIONSHIP BETWEEN DOWN-HOLE DEPTH AND TEMPERATURE
- x-axis title: DEPTH (meters)
- y-axis title: TEMPERATURE (degrees centigrade)
- re-size the chart so that it is 7 columns wide and 25 rows long
- delete the legend
- delete the gridlines
- move the chart below the table

(b) Create the *least-squares regression line* for these data on the scatterplot.
(c) Use Excel to run the regression statistics to find the *equation for the least-squares regression line* for these data and display the results below the chart on your spreadsheet. Use number format (two decimal places) for the correlation, r, and for both the y-intercept and the slope of the line. Change all other decimal figures to four decimal places.
(d) Print the input data and the chart so that this information fits onto one page.
(e) Then, print out the regression output table so that this information fits onto a separate page.

   By hand:

   (1a) Circle and label the value of the *y-intercept* and the *slope* of the regression line onto that separate page.
   (2b) *Read from the graph* the temperature you would predict for a *depth of three meters* and write your answer in the space immediately below:

(f) save the file as: DEPTH3

Answer the following questions using your Excel printout:

1. What is the correlation?
2. What is the y-intercept?
3. What is the slope of the line?
4. What is the regression equation for these data (use two decimal places for the y-intercept and the slope)?
5. Use that regression equation to predict the temperature you would expect for a down-hole depth of two meters.

   (Note that this correlation is not the multiple correlation as the Excel table indicates, but is merely the correlation r instead).
   Note that you found a positive correlation of + .94 between depth and temperature. You know that the correlation is a positive correlation for two reasons: (1) the regression line slopes upward and to the right on the chart, signaling a positive correlation, and (2) the slope is + 0.53 which also tells you that the correlation is a positive correlation.
   But how does Excel treat *negative correlations*?

Important note: Since Excel does not recognize negative correlations in the SUMMARY OUTPUT but treats all correlations as if they were positive correlations, you need to be careful to note when there is a negative correlation between the two variables under study.
You know that the correlation is negative when:

(1) *The slope, b, is a negative number which can only occur when there is a negative correlation.*
(2) *The chart clearly shows a downward slope in the regression line, which can only happen when the correlation is negative.*

3. Suppose that you wanted to study the relationship between glue strength and temperature in wooden joints using a controlled laboratory setting. The strength of the wood joints is important for many applications such as furniture, building construction, and some marine applications. Suppose that you were hired by a company to test the strength of the adhesive (glue) that they use to bond wood together under different temperatures. For the purposes of this study, you will be testing cross grain joints on small test pieces of wood. All of the wood pieces used in the test are from the same type of wood. The test pieces are all the same size, and equal amounts of glue were evenly applied to all surfaces of the wood in the joint. You decide to test the glue by applying pressure to the wooden joints using a machine that measures force in Newtons (N). In your laboratory, you are also able to control the temperature of the testing area. You want to test your Excel skills on a random sample of hypothetical data to make sure that you can do this type of research. The hypothetical data appear in Fig. 6.32:

Research question: "What is the effect of temperature on glue in wooden cross grain joints?"

Force (N)	Temperature ($^\circ$C)
600.5	0
587.2	0
622.8	5
587.2	10
653.9	12
689.5	16
751.7	19
716.2	22
595.3	26
616.7	27
796.2	29
822.9	28
680.6	30
867.4	33
898.5	36
822.9	38
978.6	38

Fig. 6.32 Worksheet Data for Chapter 6: Practice Problem #3

Create an Excel spreadsheet and enter the data using Force (N) as the independent variable (predictor) and Temperature (degrees Centigrade) as the dependent variable (criterion). Underneath the table, use Excel's =correl function to find the correlation between these two variables. Label the correlation and place it underneath the table; then round off the correlation to two decimal places.

(a) create an *XY scatterplot* of these two sets of data such that:

- top title: RELATIONSHIP BETWEEN GLUE STRENGTH AND TEMPERATURE
- x-axis title: FORCE (N)
- y-axis title: TEMPERATURE (degrees C)
- move the chart below the table
- re-size the chart so that it is 8 columns wide and 25 rows long
- delete the legend
- delete the gridlines

(b) Create the *least-squares regression line* for these data on the scatterplot, and add the regression equation to the chart.

(c) Use Excel to run the regression statistics to find the *equation for the least-squares regression line* for these data and display the results below the chart on your spreadsheet. Use number format (2 decimal places) for the correlation, and three decimal places for all other decimal figures, including the coefficients.

(d) Print just the input data and the chart so that this information fits onto one page. Then, print the regression output table on a separate page so that it fits onto that separate page.

(e) save the file as: GLUE3

Answer the following questions using your Excel printout:

1. What is the correlation between Force and Temperature?
2. What is the y-intercept?
3. What is the slope of the line?
4. What is the regression equation?
5. Use the regression equation to predict the Temperature you would expect for a force of 800 N. Show your work on a separate sheet of paper.

References

Black K. Business statistics: for contemporary decision making. 6th ed. Hoboken: John Wiley & Sons, Inc.; 2010.

Ledholter R, Hogg R. Applied statistics for engineers and physical scientists. 3rd ed. Upper Saddle River: Pearson Prentice Hall; 2010.

Levine D, Stephan D, Krehbiel T, Berenson M.Statistics for managers using microsoft excel. 6th ed.Boston: Prentice Hall Pearson; 2011.

McCleery R, Watt T, Hart T. Introduction to statistics for biology. 3rd ed. Boca Raton: Chapman & Hall/CRC; 2007.

McKillup S, Dyar M. Geostatistics explained: an introductory guide for earth scientists. Cambridge: Cambridge University Press; 2010.

Chapter 7
Multiple Correlation and Multiple Regression

There are many times in the physical sciences when you want to predict a criterion, Y, but you want to find out if you can develop a better prediction model by using *several predictors* in combination (e.g. X_1, X_2, X_3, etc.) instead of a single predictor, X.

The resulting statistical procedure is called "multiple correlation" because it uses two or more predictors in combination to predict Y, instead of a single predictor, X. Each predictor is "weighted" differently based on its separate correlation with Y and its correlation with the other predictors. The job of multiple correlation is to produce a regression equation that will weight each predictor differently and in such a way that the combination of predictors does a better job of predicting Y than any single predictor by itself. We will call the multiple correlation: R_{xy}.

You will recall (see Sect. 6.5.3) that the regression equation that predicts Y when only one predictor, X, is used is:

$$Y = a + bX \tag{7.1}$$

7.1 Multiple Regression Equation

The multiple regression equation follows a similar format and is:

$$
\begin{aligned}
Y = a &+ b_1X_1 + b_2X_2 + b_3X_3 \\
&+ etc.\,depending\,on\,the\,number\,of\,predictors\,used
\end{aligned}
\tag{7.2}
$$

The "weight" given to each predictor in the equation is represented by the letter "b" with a subscript to correspond to the same subscript on the predictors.

Important note: In order to do multiple regression, you need to have installed the "Data Analysis ToolPak" that was described in Chapter 6 (see Sect. 6.5.1). If you did not install this, you need to do so now.

T.J. Quirk et al., *Excel 2010 for Physical Sciences Statistics: A Guide to Solving Practical Problems*, DOI 10.1007/978-3-319-00630-7_7, © Springer International Publishing Switzerland 2013

Let's try a practice problem.

Suppose that you have been asked to analyze some data from the SAT Reasoning Test (formerly called the Scholastic Aptitude Test) which is a standardized test for college admissions in the U.S. This test is intended to measure a student's readiness for academic work in college, and about 1.4 million high school students take this test every year. There are three subtest scores generated from this test: Critical Reading, Writing, and Mathematics, and each of these subtests has a score range between 200–800 with an average score of about 500.

Suppose that a nearby selective college in the northeast of the U.S. that is near to you wants to determine the relationship between SAT Reading scores, SAT Writing scores, and SAT Math scores in their ability to predict freshman grade-point average (FROSH GPA) for Chemistry majors at the end of freshman year at this college, and that this college has asked you to determine this relationship.

You have decided to use the three subtest scores as the predictors, X_1, X_2, and X_3 and the freshman grade-point average (FROSH GPA) as the criterion, Y. To test your Excel skills, you have selected 11 chemistry majors randomly from last year's freshmen class, and have recorded their scores on these variables.

Let's use the following notation:

Y   FROSH GPA
X_1 READING SCORE
X_2 WRITING SCORE
X_3 MATH SCORE

Suppose, further, that you have collected the following hypothetical data summarizing these scores (see Fig. 7.1):

	A	B	C	D	E
1					
2	**SAT REASONING TEST**				
3					
4	Is there a relationahip between SAT scores and Freshman GPA at a local college?				
5					
6	**FROSH GPA**	**READING SCORE**	**WRITING SCORE**	**MATH SCORE**	
7	2.55	250	230	220	
8	3.05	610	240	440	
9	3.55	620	540	530	
10	2.05	420	420	260	
11	2.45	320	520	320	
12	2.95	630	620	620	
13	3.15	650	540	530	
14	3.45	520	580	560	
15	3.30	420	490	630	
16	2.75	330	220	610	
17	3.65	440	570	660	
18					
19					

Fig. 7.1 Worksheet Data for SAT versus FROSH GPA (Practical Example)

Create an Excel spreadsheet for these data using the following cell reference:

A2:   SAT REASONING TEST
A4:   Is there a relationship between SAT scores and Freshman GPA at a local
        college?
A6:   FROSH GPA
A7:   2.55
B6:   READING SCORE
C6:   WRITING SCORE
D6:   MATH SCORE
D17:   660

Next, change the column width to match the above table, and change all GPA figures to number format (two decimal places).

Now, fill in the additional data in the chart such that:

A17:   3.65
B17:   440
C17   570

Then, center all numbers in your table

Important note: Be sure to double-check all of your numbers in your table to be sure that they are correct, or your spreadsheets will be incorrect.

Save this file as: GPA25

Before we do the multiple regression analysis, we need to try to make one important point very clear:

Important note: When we used one predictor, X, to predict one criterion, Y, we said that you need to make sure that the X variable is ON THE LEFT in your table, and the Y variable is ON THE RIGHT in your table so that you don't get these variables mixed up (see Sect. 6.3).

However, in multiple regression, you need to follow this rule which is exactly the opposite:

When you use several predictors in multiple regression, it is essential that the criterion you are trying to predict, Y, be ON THE FAR LEFT, and all of the predictors are TO THE RIGHT of the criterion, Y, in your table so that you know which variable is the criterion, Y, and which variables are the predictors. If you make this a habit, you will save yourself a lot of grief.

Notice in the table above, that the criterion Y (FROSH GPA) is on the far left of the table, and the three predictors (READING SCORE, WRITING SCORE, and MATH SCORE) are to the right of the criterion variable. If you follow this rule, you will be less likely to make a mistake in this type of analysis.

7.2   Finding the Multiple Correlation and the Multiple Regression Equation

> Objective: To find the multiple correlation and multiple regression equation using Excel.

You do this by the following commands:

Data

Click on: Data Analysis (far right top of screen)

Regression (scroll down to this in the box; see Fig. 7.2)

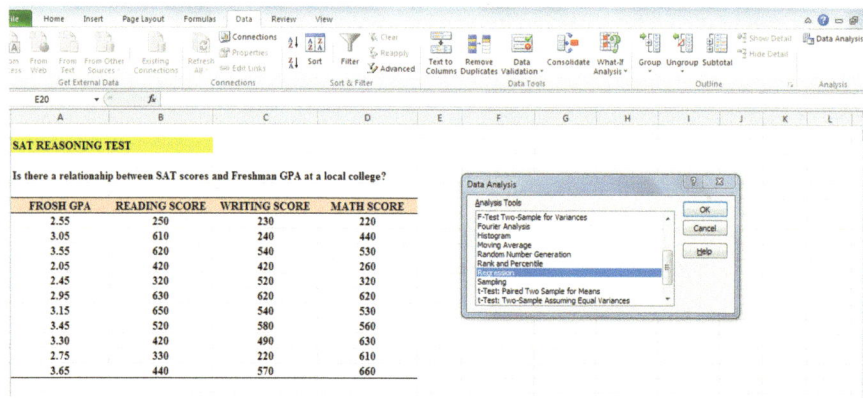

Fig. 7.2 Dialogue Box for Regression Function

OK

Input Y Range: A6:A17

Input X Range: B6:D17

Note that both the input Y Range and the Input X Range above both include the label at the top of the columns.

Click on the Labels box to *add a check mark* to it (because you have included the column labels in row 6)

Output Range (click on the button to its left, and enter): A20 (see Fig. 7.3)

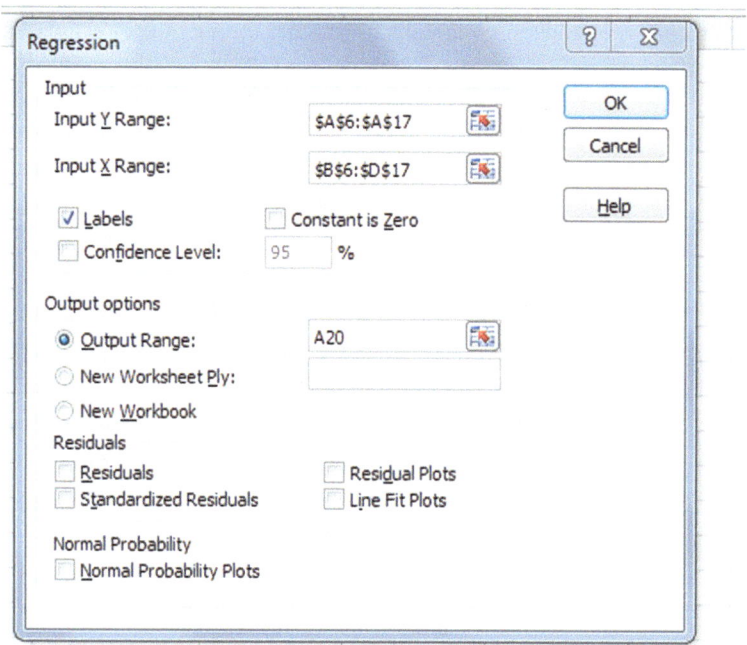

Fig. 7.3 Dialogue Box for SAT vs. FROSH GPA Data

*Important note: Excel automatically assigns a dollar sign $ in front of each column
letter and each row number so that you can keep these ranges of
data constant for the regression analysis.*

OK (see Fig. 7.4 to see the resulting SUMMARY OUTPUT)

	A	B	C	D	E	F	G	H	I	J
17	3.65	440	570	660						
18										
19										
20	SUMMARY OUTPUT									
21										
22	*Regression Statistics*									
23	Multiple R	0.797651156								
24	R Square	0.636247366								
25	Adjusted R Square	0.48035338								
26	Standard Error	0.361446932								
27	Observations	11								
28										
29	ANOVA									
30		*df*	*SS*	*MS*	*F*	*Significance F*				
31	Regression	3	1.599583719	0.533194573	4.081282	0.057174747				
32	Residual	7	0.91450719	0.130643884						
33	Total	10	2.514090909							
34										
35		*Coefficients*	*Standard Error*	*t Stat*	*P-value*	*Lower 95%*	*Upper 95%*	*Lower 95.0%*	*Upper 95.0%*	
36	Intercept	1.53627108	0.468442063	3.279532734	0.013496	0.428581617	2.643960543	0.428581617	2.643960543	
37	READING SCORE	0.000642945	0.000963026	0.667629207	0.525762	-0.001634251	0.00292014	-0.001634251	0.00292014	
38	WRITING SCORE	0.000264354	0.000889915	0.297055329	0.775046	-0.00183996	0.002368667	-0.00183996	0.002368667	
39	MATH SCORE	0.00210733	0.000848684	2.4830572	0.042022	0.000100512	0.004114149	0.000100512	0.004114149	
40										

Fig. 7.4 Regression SUMMARY OUTPUT of SAT vs. FROSH GPA Data

Next, format cell B23 in number format (2 decimal places)

Next, format the following four cells in Number format (4 decimal places):

B36

B37

B38

B39

Change all other decimal figures to two decimal places, and center all figures within their cells.

Save the file as: GPA26

Now, print the file so that it fits onto one page by changing the scale to *60 % size*. The resulting regression analysis is given in Fig. 7.5.

SAT REASONING TEST

Is there a relationahip between SAT scores and Freshman GPA at a local college?

FROSH GPA	READING SCORE	WRITING SCORE	MATH SCORE
2.55	250	230	220
3.05	610	240	440
3.55	620	540	530
2.05	420	420	260
2.45	320	520	320
2.95	630	620	620
3.15	650	540	530
3.45	520	580	560
3.30	420	490	630
2.75	330	220	610
3.65	440	570	660

SUMMARY OUTPUT

Regression Statistics	
Multiple R	0.80
R Square	0.64
Adjusted R Square	0.48
Standard Error	0.36
Observations	11

ANOVA

	df	SS	MS	F	Significance F
Regression	3	1.60	0.53	4.08	0.06
Residual	7	0.91	0.13		
Total	10	2.51			

	Coefficients	Standard Error	t Stat	P-value	Lower 95%	Upper 95%	Lower 95.0%	Upper 95.0%
Intercept	1.5363	0.47	3.28	0.01	0.43	2.64	0.43	2.64
READING SCORE	0.0006	0.00	0.67	0.53	0.00	0.00	0.00	0.00
WRITING SCORE	0.0003	0.00	0.30	0.78	0.00	0.00	0.00	0.00
MATH SCORE	0.0021	0.00	2.48	0.04	0.00	0.00	0.00	0.00

Fig. 7.5 Final Spreadsheet for SAT vs. FROSH GPA Regression Analysis

Once you have the SUMMARY OUTPUT, you can determine the multiple correlation and the regression equation that is the best-fit line through the data points using READING SCORE, WRITING SCORE, AND MATH SCORE as the three predictors, and FROSH GPA as the criterion.

Note on the SUMMARY OUTPUT where it says: "Multiple R." This term is correct since this is the term Excel uses for the multiple correlation, which is +0.80. This means, that from these data, that the combination of READING SCORES, WRITING SCORES, AND MATH SCORES together form a very strong positive relationship in predicting FROSH GPA.

To find the regression equation, *notice the coefficients at the bottom of the SUMMARY OUTPUT*:

Intercept : a (this is the y-intercept)   1.5363
READING SCORE: b_1 0.0006
WRITING SCORE: b_2 0.0003
MATH SCORE: b_3 0.0021

Since the general form of the multiple regression equation is:

$$Y = a + b_1X_1 + b_2X_2 + b_3X_3 \tag{7.2}$$

we can now write the multiple regression equation for these data:

$$Y = 1.5363 + 0.0006\,X_1 + 0.0003\,X_2 + 0.0021X_3$$

7.3 Using the Regression Equation to Predict FROSH GPA

Objective: To find the predicted FROSH GPA using an SAT Reading Score of 600, an SAT Writing Score of 500, and an SAT Math Score of 550

Plugging these three numbers into our regression equation gives us:

$$Y = 1.5363 + 0.0006\,(600) + 0.0003\,(500) + 0.0021\,(550)$$

$$Y = 1.5363 + 0.36 + 0.15 + 1.155$$

$$Y = 3.20 \text{ (since GPA scores are typically measured in two decimals)}$$

If you want to learn more about the theory behind multiple regression, see Keller (2009) and Ledolter and Hogg (2010).

7.4   Using Excel to Create a Correlation Matrix in Multiple Regression

The final step in multiple regression is to find the correlation between all of the variables that appear in the regression equation.

In our example, this means that we need to find the correlation between each of the six pairs of variables:

To do this, we need to use Excel to create a "correlation matrix." This matrix summarizes the correlations between all of the variables in the problem.

Objective: To use Excel to create a correlation matrix between the four variables in this example.

To use Excel to do this, use these steps:

Data (top of screen under "Home" at the top left of screen)
Data Analysis
Correlation (scroll *up* to highlight this formula; see Fig. 7.6)

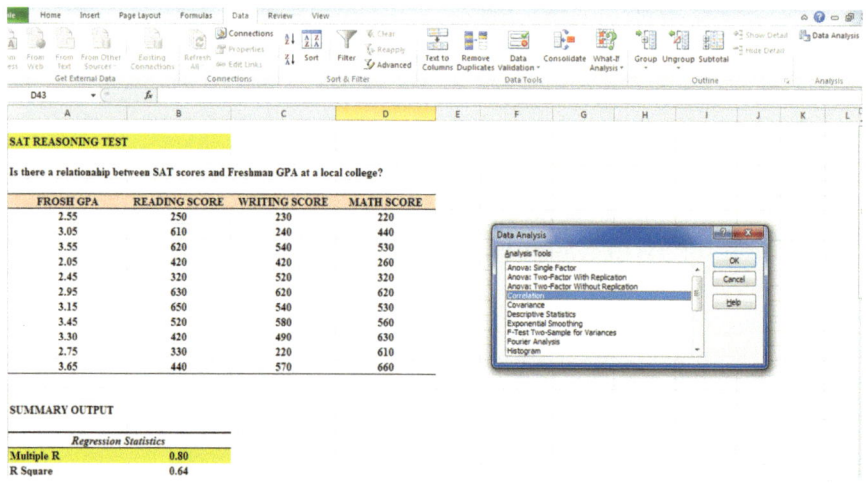

Fig. 7.6 Dialogue Box for SAT vs. FROSH GPA Correlations

OK
Input range: A6:D17

(Note that this input range includes the labels at the top of the FOUR variables (FROSH GPA, READING SCORE, WRITING SCORE, MATH SCORE) as well as all of the figures in the original data set.)

Grouped by: Columns

Put a check in the box for: Labels in the First Row (since you included the labels
    at the top of the columns in your input range of data above)
Output range (click on the button to its left, and enter):   A42 (see Fig. 7.7)

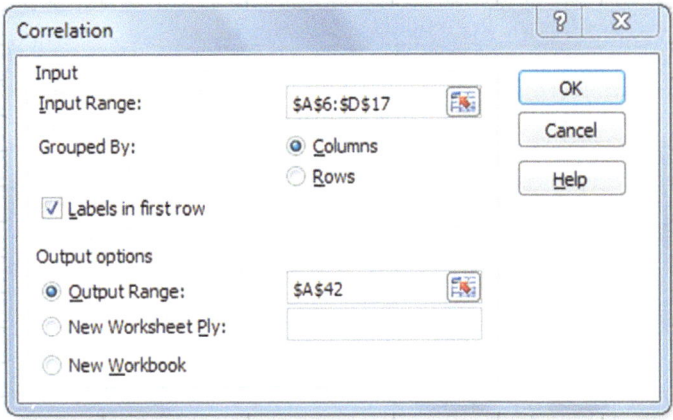

Fig. 7.7   Dialogue Box for Input / Output Range for Correlation Matrix

OK

The resulting correlation matrix appears in A42:E46 (See Fig. 7.8).

	FROSH GPA	READING SCORE	WRITING SCORE	1ATH SCORE
40				
41				
42				
43 FROSH GPA	1			
44 READING SCORE	0.510369686	1		
45 WRITING SCORE	0.446857676	0.468105152	1	
46 MATH SCORE	0.772523347	0.444074496	0.429202393	1
47				

Fig. 7.8   Resulting Correlation Matrix for SAT Scores vs. FROSH GPA Data

Next, format all of the numbers in the correlation matrix that are in decimals to
two decimals places. And, also, make column E wider so that the MATH SCORE
label fits inside cell E42.

Save this Excel file as: GPA27

The final spreadsheet for these scores appears in Fig. 7.9.

Is there a relationahip between SAT scores and Freshman GPA at a local college?

FROSH GPA	READING SCORE	WRITING SCORE	MATH SCORE
2.55	250	230	220
3.05	610	240	440
3.55	620	540	530
2.05	420	420	260
2.45	320	520	320
2.95	630	620	620
3.15	650	540	530
3.45	520	580	560
3.30	420	490	630
2.75	330	220	610
3.65	440	570	660

SUMMARY OUTPUT

Regression Statistics	
Multiple R	0.80
R Square	0.64
Adjusted R Square	0.48
Standard Error	0.36
Observations	11

ANOVA

	df	SS	MS	F	Significance F
Regression	3	1.60	0.53	4.08	0.06
Residual	7	0.91	0.13		
Total	10	2.51			

	Coefficients	Standard Error	t Stat	P-value	Lower 95%	Upper 95%	Lower 95.0%	Upper 95.0%
Intercept	1.5363	0.47	3.28	0.01	0.43	2.64	0.43	2.64
READING SCORE	0.0006	0.00	0.67	0.53	0.00	0.00	0.00	0.00
WRITING SCORE	0.0003	0.00	0.30	0.78	0.00	0.00	0.00	0.00
MATH SCORE	0.0021	0.00	2.48	0.04	0.00	0.00	0.00	0.00

	FROSH GPA	READING SCORE	WRITING SCORE	MATH SCORE
FROSH GPA	1			
READING SCORE	0.51	1		
WRITING SCORE	0.45	0.47	1	
MATH SCORE	0.77	0.44	0.43	1

Fig. 7.9 Final Spreadsheet for SAT Scores vs. FROSH GPA Regression and the Correlation Matrix

Note that the number "1" along the diagonal of the correlation matrix means that the correlation of each variable with itself is a perfect, positive correlation of 1.0.

Correlation coefficients are always expressed in just two decimal places.

You are now ready to read the correlation between the six pairs of variables:

The correlation between READING SCORE and FROSH GPA is: *+.51*
The correlation between WRITING SCORE and FROSH GPA is: *+.45*
The correlation between MATH SCORE and FROSH GPA is: *+.77*
The correlation between WRITING SCORE and READING SCORE is: *+.47*
The correlation between MATH SCORE and READING SCORE is: *+.44*
The correlation between MATH SCORE and WRITING SCORE is: *+.43*

This means that the best predictor of FROSH GPA is the MATH SCORE with a correlation of +.77. Adding the other two predictor variables, READING SCORE and WRITING SCORE, improved the prediction by only 0.03 to 0.80, and was, therefore, only slightly better in prediction. MATH SCORES are an excellent predictor of FROSH GPA all by themselves.

If you want to learn more about the correlation matrix, see Levine et al. (2011).

7.5   End-of-Chapter Practice Problems

1. The combustion rate of gunpowder varies significantly based on various chemical and physical properties. For instance, the particle size of the gunpowder used can significantly influence the rate at which the gunpowder burns. Generally speaking, the smaller the particle, the faster it burns. Faster burning particles will generate more pressure and higher temperatures. Gunpowder manufacturers will produce gunpowder with different sizes and shapes so they can control the rate at which the powder burns. The different rates of powder burn will be used for specific types of ammunitions and firearms.

   For example, gunpowder for cannons is generally larger and burns slower compared to rifle gunpowder that tends to be smaller and burn faster. If gunpowder burns too hot or too fast, it could produce too much pressure and cause the weapon's barrel to fail, and possibly explode.

   The SI unit used to measure pressure is called a "pascal" (Pa). Typical pressures in a .50 caliber military rifle are around 370 Mpa (where 1 Mpa = 1,000,000 Pa). Suppose that you were hired by a gunpowder manufacturing company to develop an experimental gunpowder load based on different gunpowder particles for use in .50 caliber military rifles. The branch of the military your company is contracted with wants a powder that has 378 Mpa of pressure at the chamber of the barrel. The chamber is part of the rifle that holds the cartridge with the bullet and gunpowder. You decide to use only one chemical formula for the gunpowder with four different particle sizes. Let's call those sizes: (1) Round, (2) Cylinder, (3) Flake, and (4) Irregular. You have also determined that a mix of the four types of powder gives the most consistent burn. This allows you to get similar results each time. The amount of gunpowder placed inside the cartridge is measured in grains (gr). A total of 214 grains (gr) will be used in all .50 caliber cartridges in the test.

   You have decided to use a multiple correlation and multiple regression analysis. To test your Excel skills, you have collected the data from a random sample of 12 test firings of the mixed powders. These hypothetical data appear in Fig. 7.10:

Research question:	"How well does gunpowder particle size predict the breech pressure in a .50 caliber rifle?"			
Breech pressure (Mpa)	Round (gr)	Cylinder (gr)	Flake (gr)	Irregular (gr)
365	73	53	44	44
375	72	52	44	46
380	75	51	38	50
375	55	75	37	47
370	52	72	41	49
370	53	71	42	48
375	47	48	44	75
365	43	43	51	77
360	51	54	38	71
355	53	50	71	40
360	44	43	73	54
360	45	48	79	42

Fig. 7.10 Worksheet Data for Chapter 7: Practice Problem #1

(a) Create an Excel spreadsheet using Breech pressure (Mpa) as the criterion (Y), and the other variables as the predictors.
(b) Use Excel's *multiple regression* function to find the relationship between these five variables and place it below the table.
(c) Use number format (2 decimal places for the multiple correlation on the SUMMARY OUTPUT, and use four decimal places for the coefficients in the SUMMARY OUTPUT).
(d) Print the table and regression results below the table so that they fit onto one page.
(e) Save this file as: GUNPOWDER8

Answer the following questions using your Excel printout:

1. What is the multiple correlation R_{xy} ?
2. What is the y-intercept a?
3. What is the coefficient for Round, b_1 ?
4. What is the coefficient for Cylinder, b_2 ?
5. What is the coefficient for Flake, b_3 ?
6. What is the coefficient for Irregular, b_4 ?
7. What is the multiple regression equation?
8. Predict the Breech pressure you would expect for a Round score of 63, a Cylinder score of 58, a Flake score of 41, and an Irregular score of 50.

(f) Now, go back to your Excel file and create a *correlation matrix* for these five variables, and place it underneath the SUMMARY OUTPUT.
(g) Re-save this file as: GUNPOWDER8
(h) Now, print out *just this correlation matrix* on a separate sheet of paper.

Answer the following questions using your Excel printout. Be sure to include the plus or minus sign for each correlation:

9. What is the correlation between Round and Breech pressure?
10. What is the correlation between Cylinder and Breech pressure?
11. What is the correlation between Flake and Breech pressure?
12. What is the correlation between Irregular and Breech pressure?
13. What is the correlation between Cylinder and Round?
14. What is the correlation between Flake and Cylinder?
15. Discuss which of the four predictors is the best predictor of Breech pressure.
16. Explain in words how much better the four predictor variables together predict Breech pressure than the best single predictor by itself.

2. Suppose you wanted to study locomotive engines for a train and that you wanted to find the relationship between the stopping distance of the train (feet), the speed at which the brakes of the train were applied which is called the enforcement speed in miles per hour (mph), and the weight of the train (tons). Let stopping distance be the dependent variable (criterion), and enforcement speed and weight be the independent variables (predictors). Hypothetical data for 13 test runs are presented in Fig. 7.11.

STOPPING DISTANCE (feet)	ENFORCEMENT SPEED (mph)	WEIGHT (tons)
45	2.9	3300
65	2.6	3800
85	4.2	3600
95	7.2	3750
100	8.6	3850
125	4.6	3950
135	12.2	4320
165	13.2	3850
186	6.5	3600
286	11.1	4750
320	15.6	3850
586	20.1	5800
650	26.2	5500

Fig. 7.11 Worksheet Data for Chapter 7: Practice Problem #2

(a) create an Excel spreadsheet using stopping distance as the criterion (Y), and the other variables as the two predictors of this criterion.
(b) Use Excel's *multiple regression* function to find the relationship between these variables and place it below the table.

(c) Use number format (2 decimal places) for the multiple correlation on the Summary Output, and use number format (three decimal places) for the coefficients and all other decimal figures in the Summary Output.
(d) Print the table and regression results below the table so that they fit onto one page.
(e) By hand on this printout, *circle and label:*

(1a) multiple correlation R_{xy}
(2b) coefficients for the y-intercept, enforcement speed, and weight.

(f) Save this file as: TRAIN3
(g) Now, go back to your Excel file and create a correlation matrix for these three variables, and place it underneath the Summary Table. *Change each correlation to just two decimals.* Save this file again as: TRAIN3
(h) Now, print out *just this correlation matrix in portrait mode* on a separate sheet of paper.

Answer the following questions using your Excel printout:

1. What is the multiple correlation R_{xy} ?
2. What is the y-intercept a ?
3. What is the coefficient for enforcement speed b_1 ?
4. What is the coefficient for weight b_2 ?
5. What is the multiple regression equation?
6. Underneath this regression equation by hand, predict the stopping distance you would expect for an enforcement speed of 10.8 mph and a weight of 4600 tons.

Answer the following questions using your Excel printout. Be sure to include the plus or minus sign for each correlation:

7. What is the correlation between enforcement speed and stopping distance?
8. What is the correlation between weight and stopping distance?
9. What is the correlation between enforcement speed and weight?
10. Discuss which of the two predictors is the better predictor of stopping distance.
11. Explain in words how much better the two predictor variables combined predict stopping distance than the better single predictor by itself.

3. There are popular home science kits known as 'Grow Your Own Crystals." These kits allow people to grow their own colorful crystals out of simple solutions. Crystal growth in these kits is usually dependent upon temperature, humidity, and time. Suppose that you are part of a new company that sells "Grow Your Own" crystal kits and that your company wants to find the best combination of temperature, humidity, and time to put in the instruction manual that will produce the most crystals if the buyer follows the directions accurately. For this example, Crystal growth (measured in centimeters, cm) is the dependent variable (criterion). Temperature (° C), Humidity (%), and Time (days) will be the independent variables (predictors).

You have decided to use a multiple correlation and multiple regression analysis, and to test your Excel skills, you have collected the data of a random sample of 10 random trials of crystal growth to collect the hypothetical data that appear in Fig. 7.12.

CRYSTAL GROWTH RATES

GROWTH (cm)	Temperature ($^{\circ}$C)	Humidity (%)	Time (days)
2.50	20	30	4
3.00	25	40	7
3.50	22	32	10
3.00	26	36	4
2.40	25	34	6
4.50	23	40	7
3.50	27	42	9
2.50	29	36	7
4.25	28	40	12
5.00	31	43	10

Fig. 7.12 Worksheet Data for Chapter 7: Practice Problem #3

(a) create an Excel spreadsheet using GROWTH (cm) as the criterion and the other three variables as the predictors.
(b) Use Excel's *multiple regression* function to find the relationship between these four variables and place the SUMMARY OUTPUT below the table.
(c) Use number format (2 decimal places) for the multiple correlation on the Summary Output, and use number format (3 decimal places) for the coefficients in the summary output and for all other decimal figures in the SUMMARY OUTPUT.
(d) Save the file as: CRYSTAL8
(e) Print the table and regression results below the table so that they fit onto one page.

Answer the following questions using your Excel printout:

1. What is multiple correlation R_{xy} ?
2. What is the y-intercept a ?
3. What is the coefficient for Temperature, b_1 ?
4. What is the coefficient for Humidity, b_2 ?
5. What is the coefficient for Time, b_3 ?
6. What is the multiple regression equation?

7. Predict the GROWTH you would expect for a Temperature of 25 degrees C, a Humidity of 34 %, and a Time of 6 days.

(f) Now, go back to your Excel file and create a correlation matrix for these four variables, and place it underneath the SUMMARY OUTPUT on your spreadsheet.
(g) Re-save this file as: CRYSTAL8
(h) Now, print out *just this correlation matrix* on a separate sheet of paper.

Answer the following questions using your Excel printout. Be sure to include the plus or minus sign for each correlation:

8. What is the correlation between Temperature and GROWTH?
9. What is the correlation between Humidity and GROWTH?
10. What is the correlation between Time and GROWTH?
11. What is the correlation between Humidity and Temperature?
12. What is the correlation between Time and Temperature?
13. What is the correlation between Time and Humidity?
14. Discuss which of the three predictors is the best predictor of GROWTH.
15. Explain in words how much better the three predictor variables combined predict GROWTH than the best single predictor by itself.

References

Keller G. Statistics for management and economics. 8th ed. Mason: South-Western Cengage Learning; 2009.
Ledolter J, Hogg R. Applied statistics for engineers and physical scientists. 3rd ed. Upper Saddle River: Pearson Prentice Hall; 2010.
Levine D, Stephan D, Krehbiel T, Berenson M. Statistics for managers using Microsoft Excel. 6th ed. Boston: Pearson Prentice Hall; 2011.

Chapter 8
One-Way Analysis of Variance (ANOVA)

So far in this 2010 Excel Guide, you have learned how to use a one-group t-test to compare the sample mean to the population mean, and a two-group t-test to test for the difference between two sample means. *But what should you do when you have more than two groups and you want to determine if there is a significant difference between the means of these groups?*

The answer to this question is: *Analysis of Variance (ANOVA).*

The ANOVA test allows you to test for the difference between the means when you have *three or more groups* in your research study.

Important note: In order to do One-way Analysis of Variance, you need to have installed the "Data Analysis Toolpak" that was described in Chapter 6 (see Sect. 6.5.1). If you did not install this, you need to do that now.

Let's suppose that you were working as a research scientist for a company and that you wanted to compare your company's premium brand of tire (Brand A) against two major competitors' brands (B and C). You have set up a laboratory test of the three types of tires, and you have measured the number of simulated miles driven before the tread length reached a predetermined amount. The hypothetical results are given in Figure 8.1. Note that the data are in thousands of miles driven (1000), so, for example, 63 is really 63,000 miles.

You have been asked to analyze the data to determine if there was any significant difference in miles driven between the three brands. To test your Excel skills, you have selected a random sample of tires from each of these brands (see Fig. 8.1). Note that each brand can have a different number of tires in order for ANOVA to be used on the data. Statisticians delight in this fact by referring to this characteristic by stating that: "ANOVA is a very robust test." (Statisticians love that term!)

T.J. Quirk et al., *Excel 2010 for Physical Sciences Statistics: A Guide to Solving Practical Problems*, DOI 10.1007/978-3-319-00630-7_8,
© Springer International Publishing Switzerland 2013

(Data are in thousands of miles)

Brand A (1000 miles)	Brand B (1000 miles)	Brand C (1000 miles)
62	61	65
61	62	67
62	63	71
64	60	66
61	64	65
	59	64
	62	
	63	
	62	
	63	

Fig. 8.1 Worksheet Data for Tire Mileage Test (Practical Example)

Create an Excel spreadsheet for these data in this way:

A4:   TIRE MILEAGE TEST
A6:   (Data are in thousands of miles)
B8:   Brand A
C8:   Brand B
D8:   Brand C
B9:   62

Enter the other information into your spreadsheet table. When you have finished entering these data, the last cell on the left should have 61 in cell B13, and the last cell on the right should have 64 in cell D14. Center the numbers in each of the columns. Use number format (zero decimals) for all numbers.

Important note: Be sure to double-check all of your figures in the table to make sure that they are exactly correct or you will not be able to obtain the correct answer for this problem!

Save this file as: TIRE6

8.1   Using Excel to Perform a One-way Analysis of Variance (ANOVA)

Objective: To use Excel to perform a one-way ANOVA test

You are now ready to perform an ANOVA test on these data using the following steps:

Data (at top of screen)
Data Analysis (far right at top of screen)
Anova: Single Factor (*scroll up to this formula and highlight it*; see Fig. 8.2)

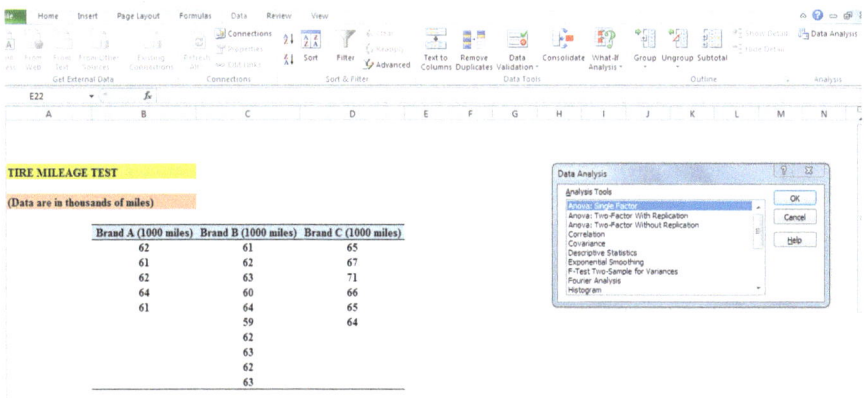

Fig. 8.2 Dialog Box for Data Analysis: Anova Single Factor

OK

Input range: B8 : D18 (note that you have included in this range the column titles that are in row 8)

Important note: Whenever the data set has a different sample size in the groups being compared, the INPUT RANGE that you define must start at the column title of the first group on the left and go to the last column on the right to the lowest row that has a figure in it in the entire data matrix so that the INPUT RANGE has the "shape" of a rectangle when you highlight it. Since Brand B has 63 in cell C18, your "rectangle" must include row 18!

Grouped by: Columns
Put a check mark in: Labels in First Row
Output range (click on the button to its left): A20 (see Fig. 8.3)

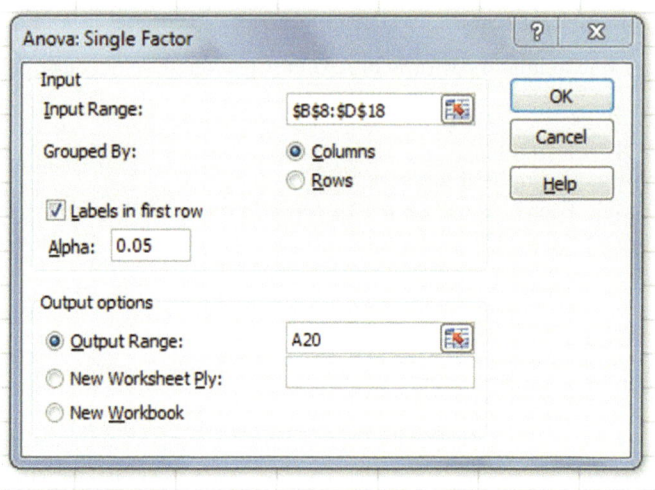

Fig. 8.3 Dialog Box for Anova: Single Factor Input / Output Range

OK

Center all of the numbers in the ANOVA table, and round off all numbers that are
    decimals to two decimal places.

Save this file as: TIRE6A

    You should have generated the table given in Fig. 8.4.

TIRE MILEAGE TEST

(Data are in thousands of miles)

Brand A (1000 miles)	Brand B (1000 miles)	Brand C (1000 miles)
62	61	65
61	62	67
62	63	71
64	60	66
61	64	65
	59	64
	62	
	63	
	62	
	63	

Anova: Single Factor

SUMMARY

Groups	Count	Sum	Average	Variance
Brand A	5	310	62.00	1.50
Brand B	10	619	61.90	2.32
Brand C	6	398	66.33	6.27

ANOVA

Source of Variation	SS	df	MS	F	P-value	F crit
Between Groups	83.00	2	41.50	12.83	0.00	3.55
Within Groups	58.23	18	3.24			
Total	141.24	20				

Fig. 8.4   ANOVA Results for Tire Mileage Test

Print out both the data table and the ANOVA summary table so that all of this information fits onto one page. (Hint: Set the Page Layout / Fit to Scale to *85 % size*).

As a check on your analysis, you should have the following in these cells:

A20:   Anova: Single Factor
D24:   62.00
D31:   41.50
E31:   12.83
G31:   3.55

Now, let's discuss how you should interpret this table:

8.2   How to Interpret the ANOVA Table Correctly

Objective: To interpret the ANOVA table correctly

ANOVA allows you to test for the differences between means when you have three or more groups of data. This ANOVA test is called the F-test statistic, and is typically identified with the letter: F.

The formula for the F-test is this:

$$F = \text{Mean Square between groups (MS}_b) \text{ divided by Mean}$$
$$\text{Square within groups (MS}_w)$$

$$F = MS_b / MS_w \tag{8.1}$$

The derivation and explanation of this formula is beyond the scope of this *Excel Guide*. In this *Excel Guide*, we are attempting to teach you *how to use Excel*, and we are not attempting to teach you the statistical theory that is behind the ANOVA formulas. For a detailed explanation of ANOVA, see Hibbert and Gooding (2006) and Black (2010).

Note that cell D31 contains $MS_b = 41.50$, while cell D32 contains $MS_w = 3.24$.

When you divide these two figures using their cell references in Excel, you get the answer for the F-test of 12.83 which is in cell E31. (Remember, Excel is more accurate than your calculator!) Let's discuss now the meaning of the figure: $F = 12.83$.

In order to determine whether this figure for F of 12.83 indicates a significant difference between the means of the three groups, the first step is to write the null hypothesis and the research hypothesis for the three brands of tires.

In our statistics mileage comparisons, the null hypothesis states that the population means of the three groups are equal, while the research hypothesis states that the population means of the three groups are not equal and that there is, therefore, a significant difference between the population means of the three groups. Which of these two hypotheses should you accept based on the ANOVA results?

8.3   Using the Decision Rule for the ANOVA F-test

To state the hypotheses, let's call Brand A as Group 1, Brand B as Group 2, and Brand C as Group 3. The hypotheses would then be:

H_0 : $\mu_1 = \mu_2 = \mu_3$
H_1 : $\mu_1 \neq \mu_2 \neq \mu_3$

The answer to this question is analogous to the decision rule used in this book for both the one-group t-test and the two-group t-test. You will recall that this rule (See Sect. 4.1.6 and Sect. 5.1.8) was:

If the absolute value of t is less than the critical t, you accept the null hypothesis.
or
If the absolute value of t is greater than the critical t, you reject the null hypothesis, and accept the research hypothesis.

Now, here is the decision rule for ANOVA:

Objective: To learn the decision rule for the ANOVA F-test

The decision rule for the ANOVA F-test is the following:

If the value for F is less than the critical F-value, accept the null hypothesis.
or
If the value of F is greater than the critical F-value, reject the null hypothesis, and accept the research hypothesis.

Note that Excel tells you the critical F-value in cell G31: 3.55
Therefore, our decision rule for the tire mileage AVOVA test is this:

Since the value of F of 12.83 is greater than the critical F-value of 3.55, we reject the null hypothesis and accept the research hypothesis.

Therefore, our conclusion, in plain English, is:

There was a significant difference between the number of miles driven between the three brands of tires.

Note that it is not necessary to take the absolute value of F of 12.83. The F-value can never be less than one, and so it can never be a negative value which requires us to take its absolute value in order to treat it as a positive value.
It is important to note that ANOVA tells us that there was a significant difference between the population means of the three groups, *but it does not tell us which pairs of groups were significantly different from each other.*

8.4   Testing the Difference Between Two Groups using the ANOVA t-test

To answer that question, we need to do a different test called the ANOVA t-test.

Objective: To test the difference between the means of two groups using an
ANOVA t-test when the ANOVA F-test results indicate a significant
difference between the population means

Since we have three groups of data (one group for each of the three brands of tires), we would have to perform three separate ANOVA t-tests to determine which pairs of groups were significantly different. This requires that we would have to perform a separate ANOVA t-test for the following pairs of groups:

(1) Brand A vs. Brand B
(2) Brand A vs. Brand C
(3) Brand B vs. Brand C

We will do just one of these pairs of tests, Brand A vs. Brand C, to illustrate the way to perform an ANOVA t-test comparing these two brands of tires. The ANOVA t-test for the other two pairs of groups would be done in the same way.

8.4.1   Comparing Brand A vs. Brand C in Miles Driven Using the ANOVA t-test

Objective: To compare Brand A vs. Brand C in miles driven using the ANOVA t-test

The first step is to write the null hypothesis and the research hypothesis for these two brands of tires.

For the ANOVA t-test, the null hypothesis is that the population means of the two groups are equal, while the research hypothesis is that the population means of the two groups are not equal (i.e., there is a significant difference between these two means). Since we are comparing Brand A (Group 1) vs. Brand C (Group 3), these hypotheses would be:

H_0 : $\mu_1 = \mu_3$
H_1 : $\mu_1 \neq \mu_3$

For Group 1 vs. Group 3, the formula for the ANOVA t-test is:

$$ANOVA\,t = \frac{\overline{X}_1 - \overline{X}_2}{s.e._{ANOVA}} \tag{8.2}$$

where

$$s.e._{ANOVA} = \sqrt{MS_w \left(\frac{1}{n_1} + \frac{1}{n_2} \right)} \tag{8.3}$$

The steps involved in computing this ANOVA t-test are:

1. Find the difference of the sample means for the two groups (62 − 66.33 = − 4.33).
2. Find $1/n_1 + 1/n_3$ (since both groups have a different number of tires in them, this becomes: $1/5 + 1/6 = 0.20 + 0.17 = 0.37$
3. Multiply MS_w times the answer for step 2 ($3.24 \times 0.37 = 1.20$)
4. Take the square root of step 3 (SQRT (1.20) = 1.09)
5. Divide Step 1 by Step 4 to find ANOVA t (− 4.33 / 1.09 = −3.97)

Note: Since Excel computes all calculations to 16 decimal places, when you use Excel for the above computations, your answer will be − 3.98 in two decimal places, but Excel's answer will be much more accurate because it is always in 16 decimal places in its computations.

Now, what do we do with this ANOVA t-test result of −3.97? In order to interpret this value of −3.97 correctly, we need to determine the critical value of t for the ANOVA t-test. To do that, we need to find the degrees of freedom for the ANOVA t-test as follows:

8.4.1.1 Finding the Degrees of Freedom for the ANOVA t-test

Objective: To find the degrees of freedom for the ANOVA t-test

The degrees of freedom (df) for the ANOVA t-test is found as follows:

df = take the total sample size of all of the groups and subtract the number of groups in your study (n_{TOTAL} − k where k = the number of groups)

In our example, the total sample size of the three groups is 21 since there are 5 tires in Group1, 10 tires in Group 2, and 6 tires in Group 3, and since there are three groups, 21 − 3 gives a degrees of freedom for the ANOVA t-test of 18.

If you look up df = 18 in the t-table in Appendix E in the degrees of freedom column (df), which is the *second column on the left of this table*, you will find that the critical t-value is 2.101.

Important note: Be sure to use the degrees of freedom column (df) in Appendix E for the ANOVA t-test critical t value

8.4.1.2 Stating the Decision Rule for the ANOVA t-test

Objective: To learn the decision rule for the ANOVA t-test

Interpreting the result of the ANOVA t-test follows the same decision rule that we used for both the one-group t-test (see Sect. 4.1.6) and the two-group t-test (see Sect. 5.1.8):

If the absolute value of t is less than the critical value of t, we accept the null hypothesis.

or

If the absolute value of t is greater than the critical value of t, we reject the null hypothesis and accept the research hypothesis.

Since we are using a type of t-test, we need to take the absolute value of t. Since the absolute value of -3.98 is greater than the critical t-value of 2.101, we reject the null hypothesis (that the population means of the two groups are equal) and accept the research hypothesis (that the population means of the two groups are significantly different from one another).

This means that our conclusion, in plain English, is as follows:

The average tire mileage for Brand C was significantly greater than the average tire mileage for Brand A (66,000 vs. 62,000).

Note that this difference in average tire mileage of about 4,000 miles between Brand A and Brand C might not seem like much, but in practical terms, this means that the average miles driven for Brand C were 7 % higher than the average miles driven for Brand A. This, clearly, is an important difference in miles driven based on our hypothetical data.

8.4.1.3   Performing an ANOVA t-test Using Excel commands

Now, let's do these calculations for the ANOVA t-test using Excel with the file you created earlier in this chapter: TIRE6A

A37:   Brand A vs. Brand C
A39:   1/5 + 1/6
A41:   s.e. ANOVA
A43:   ANOVA t-test
B39:   =(1/5 +1/6)
B41:   =SQRT(D32*B39)
B43:   =(D24 − D26)/B41

You should now have the following results in these cells when you round off all these figures in the ANOVA t-test to two decimal points.:

B39:   0.37
B41:   1.09
B43:   −3.98

Save this final result under the file name: TIRE7

Print out the resulting spreadsheet so that it fits onto one page like Figure 8.5 (Hint: Reduce the Page Layout / Scale to Fit to 85 %).

TIRE MILEAGE TEST

(Data are in thousands of miles)

Brand A (1000 miles)	Brand B (1000 miles)	Brand C (1000 miles)
62	61	65
61	62	67
62	63	71
64	60	66
61	64	65
	59	64
	62	
	63	
	62	
	63	

Anova: Single Factor

SUMMARY

Groups	Count	Sum	Average	Variance
Brand A	5	310	62.00	1.50
Brand B	10	619	61.90	2.32
Brand C	6	398	66.33	6.27

ANOVA

Source of Variation	SS	df	MS	F	P-value	F crit
Between Groups	83.00	2	41.50	12.83	0.0003	3.55
Within Groups	58.23	18	3.24			
Total	141.24	20				

Brand A vs. Brand C

1/5 + 1/6	0.37
s.e. ANOVA	1.09
ANOVA t-test	-3.98

Fig. 8.5   Final Spreadsheet of Tire Mileage for Brand A vs. Brand C

For a more detailed explanation of the ANOVA t-test, see Black (2010).

Important note: You are only allowed to perform an ANOVA t-test comparing the means of two groups when the F-test produces a significant difference between the means of all of the groups in your study.

It is improper to do any ANOVA t-test when the value of F is less than the critical value of F. Whenever F is less than the critical F, this means that there was no difference between the means of the groups, and, therefore that you cannot test to see if there is a difference between the means of any two groups since this would capitalize on chance differences between these two groups. For more information on this important point, see Gould et al. (2002).

8.5   End-of-Chapter Practice Problems

1. Let's suppose that you have been asked to study the yield (grams of product produced) of a chemical reaction conducted under three different temperature conditions: (1) BELOW ROOM TEMPERATURE (15 degrees Celsius ($^{\circ}$C)), (2) ROOM TEMPERATURE (25 degrees Celsius ($^{\circ}$C)), and (3) ABOVE ROOM TEMPERATURE (30 degrees Celsius ($^{\circ}$C)).

   You have been asked to analyze the data from the reactions to determine if there was a significant difference in yield (grams of product produced) between the three different temperatures. To test your Excel skills, you have selected a random sample of results from each of the three chemical reactions performed at different temperatures (see Fig. 8.6). Note that each group reactions can be of a different number of results in order for ANOVA to be used on the data. Statisticians delight in this fact by referring to this characteristic by stating that: "ANOVA is a very robust test." (Statisticians love that term!)

CHEMICAL REACTION YIELD (grams of product produced)

BELOW ROOM TEMP (15 $^{\circ}$C)	ROOM TEMP (25 $^{\circ}$C)	ABOVE ROOM TEMP (30 $^{\circ}$C)
90	85	76
85	89	80
74	83	90
89	79	84
84	74	78
95	75	65
92	86	42
65	87	58
75	86	63
73	88	75
54		66
71		

Fig. 8.6 Worksheet Data for Chapter 8: Practice Problem #1

(a) Enter these data on an Excel spreadsheet.
(b) Perform a *one-way ANOVA test* on these data, and show the resulting ANOVA table *underneath* the input data for the three temperatures.
(c) If the F-value in the ANOVA table is significant, create an Excel formula to compute the ANOVA t-test comparing the average for ROOM TEMPERA-TURE against ABOVE ROOM TEMPERATURE and show the results below the ANOVA table on the spreadsheet (put the standard error and the ANOVA t-test value on separate lines of your spreadsheet, and use two decimal places for each value)
(d) Print out the resulting spreadsheet so that all of the information fits onto one page
(e) Save the spreadsheet as: REACTION3

Now, write the answers to the following questions using your Excel printout:

1. What are the null hypothesis and the research hypothesis for the ANOVA F-test?
2. What is MS_b on your Excel printout?
3. What is MS_w on your Excel printout?
4. Compute $F = MS_b / MS_w$ using your calculator.
5. What is the critical value of F on your Excel printout?
6. What is the result of the ANOVA F-test?
7. What is the conclusion of the ANOVA F-test in plain English?
8. If the ANOVA F-test produced a significant difference between the three types of temperatures, what is the null hypothesis and the research hypothesis for the ANOVA t-test comparing ROOM TEMPERATURE versus ABOVE ROOM TEMPERATURE?
9. What is the mean (average) for ROOM TEMPERATURE on your Excel printout?
10. What is the mean (average) for ABOVE ROOM TEMPERATURE on your Excel printout?
11. What are the degrees of freedom (df) for the ANOVA t-test comparing ROOM TEMPERATURE versus ABOVE ROOM TEMPERATURE?
12. What is the critical t value for this ANOVA t-test in Appendix E for these degrees of freedom?
13. Compute the $s.e._{ANOVA}$ using your calculator.
14. Compute the ANOVA t-test value comparing ROOM TEMPERATURE versus ABOVE ROOM TEMPERATURE using your calculator.
15. What is the result of the ANOVA t-test comparing ROOM TEMPERA-TURE versus ABOVE ROOM TEMPERATURE?
16. What is the conclusion of the ANOVA t-test comparing ROOM TEMPER-ATURE versus ABOVE ROOM TEMPERATURE in plain English?

Note that since there are three types of temperatures, you need to do three ANOVA t-tests to determine what the significant differences are between the three types of temperatures. *Since you have just completed the ANOVA t-test comparing ROOM TEMPERATURE versus ABOVE ROOM TEMPERATURE, you would also need to do the ANOVA t-test comparing ROOM TEMPERATURE versus BELOW ROOM TEMPERATURE, and also the ANOVA t-test comparing ABOVE ROOM TEMPERATURE versus BELOW ROOM TEMPERATURE in order to write a conclusion summarizing these three types of ANOVA t-tests.*

2. Small differences in horsepower output by race car engines can mean the difference between winning and losing professional races. Suppose that you have been hired by a sponsor to develop a new type of fuel injection system for its engine in order to increase the horsepower of the engine. You have worked with different engineers and developed four prototypes of fuel injectors. Let's call the different fuel injectors: A, B, C, and D. The four types of fuel injectors have been placed on several engines to test their horsepower output. You decide to take a random sample of horsepower output readings from the different engines to test your Excel skills, and the hypothetical data are given in Fig. 8.7.

HORSEPOWER OUTPUT WITH PROTOTYPE FUEL INJECTOR

FUEL INJECTOR A	FUEL INJECTOR B	FUEL INJECTOR C	FUEL INJECTOR D
550	550	555	555
600	560	565	570
650	580	575	580
625	600	580	575
620	610	590	585
670	630	625	590
680	660	640	610
700	670	665	630
725	690	680	625
710		690	645
		695	690
			680

Fig. 8.7   Worksheet Data for Chapter 8: Practice Problem #2

(a) Enter these data on an Excel spreadsheet.
(b) Perform a *one-way ANOVA test* on these data, and show the resulting ANOVA table *underneath* the input data for the four types of fuel injectors. Round off all decimal figures to two decimal places, and center all numbers in the ANOVA table.
(c) If the F-value in the ANOVA table is significant, create an Excel formula to compute the ANOVA t-test comparing the horsepower output for FUEL INJECTOR A against the horsepower output for FUEL INJECTOR C, and show the results below the ANOVA table on the spreadsheet (put the

standard error and the ANOVA t-test value on separate lines of your spreadsheet, and use two decimal places for each value)

(d) Print out the resulting spreadsheet so that all of the information fits onto one page

(e) Save the spreadsheet as: FUEL3

Now, write the answers to the following questions using your Excel printout:

1. What are the null hypothesis and the research hypothesis for the ANOVA F-test?
2. What is MS_b on your Excel printout?
3. What is MS_w on your Excel printout?
4. Compute $F = MS_b / MS_w$ using your calculator.
5. What is the critical value of F on your Excel printout?
6. What is the result of the ANOVA F-test?
7. What is the conclusion of the ANOVA F-test in plain English?
8. If the ANOVA F-test produced a significant difference between the four types of fuel injectors in their horsepower output, what is the null hypothesis and the research hypothesis for the ANOVA t-test comparing FUEL INJECTOR A versus FUEL INJECTOR C?
9. What is the mean (average) horsepower for FUEL INJECTOR A on your Excel printout?
10. What is the mean (average) horsepower for FUEL INJECTOR C on your Excel printout?
11. What are the degrees of freedom (df) for the ANOVA t-test comparing FUEL INJECTOR A versus FUEL INJECTOR C?
12. What is the critical t value for this ANOVA t-test in Appendix E for these degrees of freedom?
13. Compute the s.e.$_{ANOVA}$ using Excel for FUEL INJECTOR A versus FUEL INJECTOR C.
14. Compute the ANOVA t-test value comparing FUEL INJECTOR A versus FUEL INJECTOR C using Excel.
15. What is the result of the ANOVA t-test comparing FUEL INJECTOR A versus FUEL INJECTOR C?
16. What is the conclusion of the ANOVA t-test comparing FUEL INJECTOR A versus FUEL INJECTOR C in plain English?

3. Suppose that you were working as a research scientist and that you wanted to do a research study comparing the highway miles per gallon (mpg) for five types of vehicles: (1) SUBCOMPACTS, (2) COMPACTS, (3) MID-SIZE, (4) LARGE, and (5) SUVs. You want to answer the research question: Is the size of the vehicle related to gasoline usage? You have obtained the cooperation of the owners of each type of vehicle who agree to keep track of their highway mileage over a pre-determined route for three tanks of gasoline. The hypothetical data for this study are given in Fig. 8.8.

1	2	3	4	5
SUBCOMPACTS (mpg)	COMPACTS (mpg)	MID-SIZE (mpg)	LARGE (mpg)	SUVs (mpg)
28.1	26.2	24.0	22.0	18.1
30.2	28.3	26.3	23.1	20.2
29.3	29.3	25.2	25.4	22.3
31.6	27.0	27.1	24.3	21.4
33.0	28.0	28.0	25.0	20.5
34.3	29.5	23.6	24.7	19.0
32.1	31.0	29.2	23.1	18.2
35.0	32.3		22.4	19.1
	33.1		26.0	
			21.3	

Fig. 8.8 Worksheet Data for Chapter 8: Practice Problem #3

(a) Enter these data on an Excel spreadsheet.
(b) Perform a *one-way ANOVA test* on these data, and show the resulting ANOVA table *underneath* the input data for the five types of vehicles.
(c) If the F-value in the ANOVA table is significant, create an Excel formula to compute the ANOVA t-test comparing the average mpg for COMPACTS against the average mpg for LARGE, and show the results below the ANOVA table on the spreadsheet (put the standard error and the ANOVA t-test value on separate lines of your spreadsheet, and use two decimal places for each value)
(d) Print out the resulting spreadsheet so that all of the information fits onto one page
(e) Save the spreadsheet as: CARS3

Now, write the answers to the following questions using your Excel printout:

1. What are the null hypothesis and the research hypothesis for the ANOVA F-test?
2. What is MS_b on your Excel printout?
3. What is MS_w on your Excel printout?
4. Compute $F = MS_b / MS_w$ using your calculator.
5. What is the critical value of F on your Excel printout?
6. What is the result of the ANOVA F-test?
7. What is the conclusion of the ANOVA F-test in plain English?
8. If the ANOVA F-test produced a significant difference between the five types of vehicles in their mpg, what is the null hypothesis and the research hypothesis for the ANOVA t-test comparing COMPACTS versus LARGE?
9. What is the mean (average) mpg for COMPACTS on your Excel printout?
10. What is the mean (average) mpg for LARGE on your Excel printout?
11. What are the degrees of freedom (df) for the ANOVA t-test comparing COMPACTS versus LARGE?

12. What is the critical t value for this ANOVA t-test in Appendix E for these degrees of freedom?
13. Compute the s.e.$_{ANOVA}$ using your calculator for COMPACTS versus LARGE.
14. Compute the ANOVA t-test value comparing COMPACTS versus LARGE using your calculator.
15. What is the result of the ANOVA t-test comparing COMPACTS versus LARGE?
16. What is the conclusion of the ANOVA t-test comparing COMPACTS versus LARGE in plain English?

References

Black K. Business statistics: for contemporary decision making. 6th ed. Hoboken: John Wiley & Sons, Inc.; 2010.

Gould J, Gould G. Biostats basics: a student handbook. New York: W.H. Freeman and Company; 2002.

Hibbert D, Gooding J. Data analysis for chemistry: an introductory guide for students and laboratory scientists. New York: Oxford University Press; 2006.

Appendix A
Answers to End-of-Chapter Practice Problems

Chapter 1: *Practice Problem #1 Answer (see Fig. A.1)*

IRON ORE (LIMONITE) SAMPLES

Percent (%) iron		
18.24		
18.29	n	13
18.26		
18.28		
18.30	Mean	18.27
18.24		
18.26		
18.25	STDEV	0.02
18.28		
18.29		
18.30	s.e.	0.01
18.24		
18.26		

Fig. A.1 Answer to Chapter 1: Practice Problem #1

T.J. Quirk et al., *Excel 2010 for Physical Sciences Statistics: A Guide to Solving Practical Problems*, DOI 10.1007/978-3-319-00630-7,
© Springer International Publishing Switzerland 2013

Chapter 1: *Practice Problem #2 Answer (see Fig. A.2)*

LEAD CONCENTRATION IN AIR SAMPLES TAKEN NEAR SAN FRANCISCO

Micrograms per cubic meter ($\mu g/m^3$)		
3.1		
10.1	n	14
6.7		
8.9		
5.6	Mean	7.44
6.4		
4.8		
10.2	STDEV	2.25
9.8		
8.4		
7.5	s.e.	0.60
9.4		
8.5		
4.8		

Fig. A.2 Answer to Chapter 1: Practice Problem #2

Chapter 1: *Practice Problem #3 Answer (see Fig. A.3)*

SILVER ORE SAMPLES

PERCENT (%) SILVER		
12		
15	n	16
13		
8		
10	Mean	11.750
12		
13		
12	STDEV	2.910
9		
4		
11	s.e.	0.727
15		
13		
15		
12		
14		

Fig. A.3 Answer to Chapter 1: Practice Problem #3

Chapter 2: *Practice Problem #1 Answer (see Fig. A.4)*

FRAME NUMBERS	Duplicate frame numbers	RANDOM NO.
1	7	0.701
2	50	0.491
3	23	0.845
4	13	0.908
5	6	0.000
6	14	0.863
7	60	0.490
8	37	0.404
9	61	0.381
10	33	0.167
11	43	0.455
12	4	0.229
13	8	0.413
14	5	0.408
15	59	0.514
16	39	0.092
17	63	0.119
18	35	0.286
19	49	0.929
20	1	0.098
21	57	0.981
22	12	0.475
23		
	5	0.6.
56	19	0.585
57	45	0.698
58	21	0.337
59	38	0.709
60	17	0.690
61	11	0.787
62	42	0.358
63	54	0.629

Fig. A.4 Answer to Chapter 2: Practice Problem #1

Chapter 2: *Practice Problem #2 Answer (see Fig. A.5)*

Fig. A.5 Answer to
Chapter 2: Practice
Problem #2

FRAME NO.	Duplicate frame no.	Random number
1	58	0.492
2	50	0.219
3	43	0.798
4	42	0.442
5	86	0.169
6	24	0.948
7	22	0.428
8	11	0.473
9	104	0.222
10	105	0.104
11	61	0.672
12	41	0.251
13	79	0.664
14	93	0.181
15	85	0.370
16	54	0.706
17	16	0.131
18	77	0.322
19	15	0.917
20	112	0.156
21	28	0.920
22	102	0.118
3	107	

5		0.500
99	40	0.622
100	96	0.274
101	48	0.002
102	108	0.397
103	109	0.182
104	33	0.945
105	3	0.735
106	90	0.145
107	110	0.305
108	62	0.903
109	88	0.380
110	60	0.998
111	98	0.371
112	94	0.542
113	59	0.248
114	67	0.364

Chapter 2: *Practice Problem #3 Answer (see Fig. A.6)*

FRAME NUMBERS	Duplicate frame numbers	Random number
1	58	0.819
2	7	0.028
3	37	0.488
4	49	0.399
5	26	0.482
6	65	0.633
7	48	0.402
8	63	0.944
9	15	0.027
10	25	0.854
11	21	0.906
12	36	0.531
13	43	0.212
14	11	0.107
15	10	0.133
16	39	0.472
17	72	0.793
18	59	0.643
19	16	0.151
20	54	0.995
21	52	0.763
22	3	0.783
23	45	0.502
24		15
	7	0.5
70	57	0.786
71	60	0.526
72	9	0.804
73	20	0.239
74	18	0.087
75	5	0.214

Fig. A.6 Answer to Chapter 2: Practice Problem #3

Chapter 3: *Practice Problem #1 Answer (see Fig. A.7)*

EXPECTED LIFETIME OF A NEW TYPE OF PASSENGER CAR TIRE

Research question: "Does this new type of synthetic tire have an
                       expected lifetime of 40,000 miles?

LIFETIME IN MILES				
38,400				
39,500	Null hypothesis:	μ	=	40,000 miles
39,400				
42,300				
46,700	Research hypothesis:	μ	$\neq$	40,000 miles
45,800				
44,300				
38,600	n	15		
42,500				
41,600	Mean	41020.00		
40,200				
38,600	STDEV	2785.47		
37,900				
38,900	s.e.	719.21		
40,600				

95% confidence interval

        lower limit   39477.46

        upper limit   42562.54

39477 ------------ ------40000	---41020-- ------------ ------42563
lower	Ref. Mean	upper
limit	Value	limit

Result: Since the reference value is inside the confidence interval,
             we accept the null hypothesis.

Conclusion: This new type of synthetic passenger car tire does have an
             expected lifetime of 40,000 miles.

Fig. A.7 Answer to Chapter 3: Practice Problem #1

Chapter 3: *Practice Problem #2 Answer (see Fig. A.8)*

EXPECTED LIFETIME OF A NEW TYPE OF LIGHT BULB

Research question:	"Does this new type of light bulb have an expected lifetime of 1300 hours?

LIFETIME IN HOURS				
1252.3				
1310.6	Null hypothesis:	μ	=	1300 hours
1264.1				
1244.2	Research hypothesis:	μ	$\neq$	1300 hours
1282.8				
1308.4	n	16		
1319.4				
1277.4	Mean	1283.4		
1289.7				
1292.3	STDEV	22.4		
1256.2				
1288.4	s.e.	5.6		
1279.9				
1264.7				
1305.8	95% confidence interval			
1297.5				
	lower limit	1271.4		
	upper limit	1295.3		

1271.4 ---------------- ----1283.4---- ------------ ----1295.3 --------1300 --------			
lower	Mean	upper	Ref.
limit		limit	Value

Result:	Since the reference value is outside of the confidence interval, we reject the null hypothesis and accept the research hypothesis.

Conclusion:	The new type of light bulb lasts significantly less than 1300 hours, and it is probably closer to 1283 hours.

Fig. A.8 Answer to Chapter 3: Practice Problem #2

Chapter 3: *Practice Problem #3 Answer (see Fig. A.9)*

WELCH'S 100% GRAPE JUICE

Research question: "Does the average can of Welch's 100% grape juice
                        produced today contain 163 ml of grape juice?

ml				
165	Null hypothesis:	μ	=	163 ml
158				
163	Research hypothesis:	μ	$\neq$	163 ml
159				
154				
157	n	13		
159				
161	Mean	159.62		
164				
154	STDEV	3.59		
157				
161	s.e.	1.00		
163				

95% confidence interval

lower limit   157.44

upper limit   161.79

157.44 --------159.62 --------------- ---161.79 --------163 -------
lower   Mean   upper   Ref
limit   limit   Value

Result: Since the reference value is outside of the confidence interval,
            we reject the null hypothesis and accept the research hypothesis.

Conclusion: Cans of 100% grape juice produced today contained significantly
             less than 163 ml, and it was probably closer to 160 ml.

Fig. A.9 Answer to Chapter 3: Practice Problem #3

Chapter 4: *Practice Problem #1 Answer (see Fig. A.10)*

PHOSPORUS CONCENTRATION (mg/L) IN WASTE WATER EFFLUENT

CONCENTRATION (mg/L)				
0.0142	Null hypothesis:	μ	=	0.015 mg/L
0.0135				
0.0138	Research hypothesis:	μ	$\neq$	0.015 mg/L
0.0136				
0.0137	n	15		
0.0135				
0.0141				
0.0140	Mean	0.0137		
0.0138				
0.0134				
0.0135	STDEV	0.0003		
0.0137				
0.0142				
0.0132	s.e.	0.0001		
0.0133				
	critical t	2.145		
	t-test	-15.92		

Result: Since the absolute value of – 15.92 is greater than the critical t
of 2.145, we reject the null hypothesis and accept the research
hypothesis.

Conclusion: There was significantly less total phosporus concentration in the waste
water effluent produced by the chemical plant than the chemical standard
of 0.015 mg/L, and it was probably closer tp 0.0137 mg/L.

Fig. A.10 Answer to Chapter 4: Practice Problem #1

Chapter 4: *Practice Problem #2 Answer (see Fig. A.11)*

WASTE GARBABE COLLECTED IN HOUSEHOLDS

Central West End, St. Louis, Missouri (USA)

WEEKLY GARBAGE COLLECTION (kg)		
18		
21		
20	Null hypothesis:	μ = 26 kg/wk
19		
21	Research hypothesis:	μ ≠ 26 kg/wk
23		
22		
24	n	26
18		
17		
19	Mean	21.42
20		
21		
24	STDEV	2.72
26		
21		
18	s.e.	0.53
25		
19		
26	critical t	2.060
21		
23		
22	t-test	-8.59
24		
26		
19		

Result: since the absolute value of − 8.59 is greater than the critical t
of 2.060, we reject the null hypothesis and accept the research
hypothesis.

Conclusion: There was significantly less garbage collected per week after
the recycling program continued for 26 weeks than before the
recycling program was initiated, and it was probably closer
to 21 kg per week, down from 26 kg/week before the recycling
program was initiated.

Fig. A.11   Answer to Chapter 4: Practice Problem #2

Chapter 4: *Practice Problem #3 Answer (see Fig. A.12)*

DISSOLVED OXYGEN CONTENT(DO) IN MAINE LAKES

DO (mg/L)				
4.6	Null hypothesis:	μ	=	5 mg/L
4.4				
4.8	Research hypothesis:	μ	$\neq$	5 mg/L
6.4				
6.5				
6.7	n		16	
6.5				
5.6				
5.4	Mean		5.52	
5.8				
4.9				
5.2	STDEV		0.72	
5.6				
5.7				
5.4	s.e.		0.18	
4.8				
	critical t		2.131	
	t-test		2.87	

Result: Since the absolute value of 2.87 is greater than the critical t of 2.131, we reject the null hypothesis and accept the research hypothesis.

Conclusion: The average level of DO in the Maine named lakes is significantly greater than 5 mg/L, and is probably closer to 5.5 mg/L

Fig. A.12 Answer to Chapter 4: Practice Problem #3

Chapter 5: *Practice Problem #1 Answer (see Fig. A.13)*

LEAD WIRES: CURRENT WIRES VS. NEW WIRES

Number of Misfeeding Wires Per Hour

Group	n	Mean	STDEV
1 Current wires	112	21.1	3.24
2 New wires	126	19.6	3.06

Null hypothesis:	μ_1	=	μ_2
Research hypothesis:	μ_1	≠	μ_2

STDEV1 squared / n1	0.094
STDEV2 squared / n2	0.074
B17+B19	0.168
s.e.	0.410
critical t	1.960
t-test	3.659

Result: Since the absolute value of 3.659 is greater than the critical t of 1.960, we reject the null hypothesis and accept the research hypothesis.

Conclusion: The new wires had significantly fewer misfeeds per hour than the current wires (19.6 vs. 21.1)

Fig. A.13 Answer to Chapter 5: Practice Problem #1

Chapter 5: *Practice Problem #2 Answer (see Fig. A.14)*

DRYING TIME FOR HOUSEHOLD PRIMER PAINT (minutes)

Current paint	New paint
118	111
116	115
121	112
124	113
117	114
115	119
110	118
117	117
119	120
121	114
123	112
125	

Null hypothesis: μ_1 = μ_2

Research hypothesis: μ_1 ≠ μ_2

Group	n	Mean	STDEV
1 Current paint	12	118.83	4.26
2 New paint	11	115.00	3.07

(n1 - 1)x STDEV1 squared	199.67
(n2 - 1) x STDEV2 squared	94.00
n1 + n2 -2	21
1/n1 + 1/n2	0.17
s.e.	1.56
critical t	2.080
t-test	2.46

Result: Since the absolute value of 2.46 is greater than the critical t of 2.080,
               we reject the null hypothesis and accept the research hypothesis.

Conclusion: The new paint dried significantly faster than the current paint
               (115 min. vs. 119 min.)

Fig. A.14 Answer to Chapter 5: Practice Problem #2

Chapter 5: *Practice Problem #3 Answer (see Fig. A.15)*

TENSILE STRENGTH OF TWO TYPES OF HIGH CARBON STEEL USED IN FISH HOOK MANUFACTURING

PREMIERE (MPa)	STANDARD (MPa)
761	740
760	738
755	736
763	742
765	747
745	744
758	739
761	730
763	737
757	743
750	741
759	738
764	725
767	745
762	743
756	
758	

Group	n	Mean	STDEV
1 PREMIERE	17	759.06	5.49
2 STANDARD	15	739.20	5.75

Null hypothesis: μ_1 = μ_2

Research hypothesis: μ_1 ≠ μ_2

(n1 - 1) x STDEV1 squared	482.94
(n2 - 1) x STDEV2 squared	462.40
n1 + n2 - 2	30
1/n1 + 1/n2	0.13
s.e.	1.99
critical t	2.042
t-test	9.99

Result: Since the absolute value of 9.99 is greater than the critical t of 2.042,
                    we reject the null hypothesis and accept the research hypothesis.

Conclusion: The PREMIERE type of high carbon steel had significantly
                     stronger tensile strength than the STANDARD type
                     (759 MPa vs. 739 MPa).

Fig. A.15 Answer to Chapter 5: Practice Problem #3

Chapter 6: *Practice Problem #1 Answer (see Fig. A.16)*

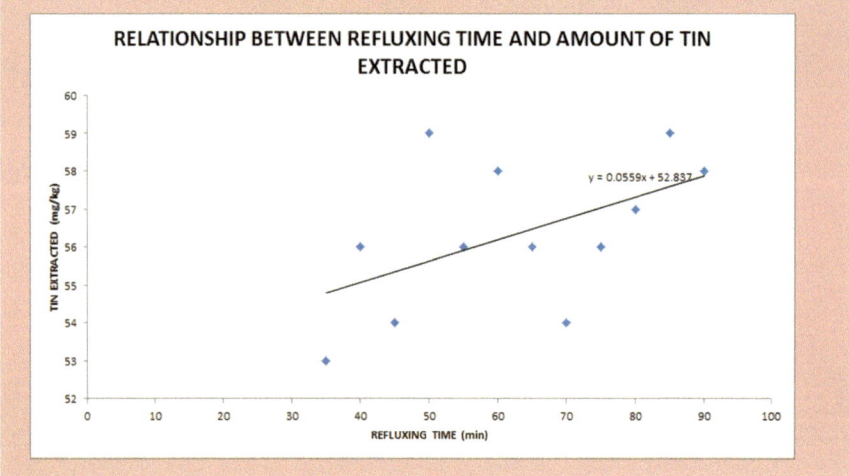

RELATIONSHIP BETWEEN REFLUXING TIME AND AMOUNT OF TIN EXTRACTED

Refluxing Time (min)	Tin Extracted (mg/kg)
35	53
40	56
45	54
50	59
55	56
60	58
65	56
70	54
75	56
80	57
85	59
90	58

correlation	0.51

RELATIONSHIP BETWEEN REFLUXING TIME AND AMOUNT OF TIN EXTRACTED

$y = 0.0559x + 52.837$

SUMMARY OUTPUT

Regression Statistics	
Multiple R	0.51
R Square	0.262
Adjusted R Square	0.188
Standard Error	1.774
Observations	12

ANOVA

	df	SS	MS	F	Significance F
Regression	1	11.189	11.189	3.555	0.089
Residual	10	31.478	3.148		
Total	11	42.667			

	Coefficients	Standard Error	t Stat	P-value	Lower 95%	Upper 95%	Lower 95.0%	Upper 95.0%
Intercept	52.837	1.924	27.462	0.000	48.550	57.124	48.550	57.124
X Variable 1	0.056	0.030	1.885	0.089	-0.010	0.122	-0.010	0.122

Fig. A.16 Answer to Chapter 6: Practice Problem #1

Chapter 6: *Practice Problem #1 (continued)*

1. $r = +.51$
2. $a = y\text{-intercept} = + 52.837$
3. $b = \text{slope} = 0.056$
4. $Y = a + b\,X$
   $Y = 52.837 + 0.056\,X$
5. $Y = 52.837 + 0.056\,(60)$
   $Y = 52.837 + 3.36$
   $Y = 56.20 \text{ mg/kg}$

Chapter 6: *Practice Problem #2 Answer (see Fig. A.17)*

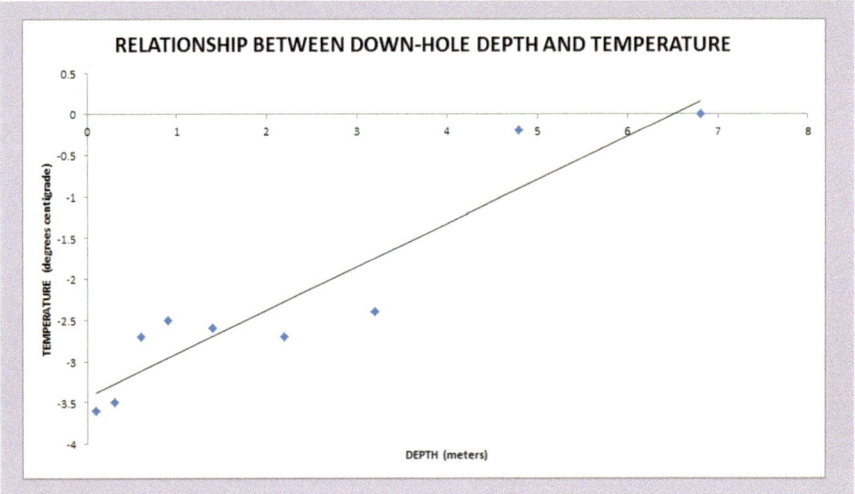

DOWN-HOLE DEPTH (meters) VS. TEMPERATURE (degrees centigrade)

DEPTH (m)	TEMPERATURE (°C)
0.1	-3.6
0.3	-3.5
0.6	-2.7
0.9	-2.5
1.4	-2.6
2.2	-2.7
3.2	-2.4
4.8	-0.2
6.8	0.0

SUMMARY OUTPUT

Regression Statistics	
Multiple R	0.94
R Square	0.8780
Adjusted R Square	0.8605
Standard Error	0.4808
Observations	9

ANOVA

	df	SS	MS	F	Significance F
Regression	1	11.6439	11.6439	50.3659	0.0002
Residual	7	1.6183	0.2312		
Total	8	13.2622			

	Coefficients	Standard Error	t Stat	P-value	Lower 95%	Upper 95%	Lower 95.0%	Upper 95.0%
Intercept	-3.43	0.2320	-14.8055	0.0000	-3.9835	-2.8863	-3.9835	-2.8863
X Variable 1	0.53	0.0744	7.0969	0.0002	0.3519	0.7036	0.3519	0.7036

Fig. A.17   Answer to Chapter 6: Practice Problem #2

Chapter 6: *Practice Problem #2 (continued)*

(2b) about -1.9 degrees centigrade

1. $r = +.94$
2. $a = $ y-intercept $= -3.43$
3. $b = $ slope $= +0.53$
4. $Y = a + bX$
   $Y = -3.43 + 0.53\ X$
5. $Y = -3.43 + 0.53\ (2)$
   $Y = -3.43 + 1.06$
   $Y = -2.37$ degrees centigrade

Chapter 6: *Practice Problem #3 Answer (see Fig. A.18)*

Research question: "What is the effect of temperature on glue in wooden cross grain joints?"

Force (N)	Temperature (°C)
600.5	0
587.2	0
622.8	5
587.2	10
653.9	12
689.5	16
751.7	19
716.2	22
595.3	26
616.7	27
796.2	29
822.9	28
680.6	30
867.4	33
898.5	36
822.9	38
978.6	38

correlation	0.78

RELATIONSHIP BETWEEN GLUE STRENGTH AND TEMPERATURE

$y = 0.0802x - 36.277$

SUMMARY OUTPUT

Regression Statistics	
Multiple R	0.78
R Square	0.603
Adjusted R Square	0.577
Standard Error	8.232
Observations	17

ANOVA

	df	SS	MS	F	Significance F
Regression	1	1546.934	1546.934	22.825	0.00024441
Residual	15	1016.596	67.773		
Total	16	2563.529			

	Coefficients	Standard Error	t Stat	P-value	Lower 95%	Upper 95%	Lower 95.0%	Upper 95.0%
Intercept	-36.277	12.300	-2.949	0.010	-62.494	-10.061	-62.494	-10.061
X Variable 1	0.080	0.017	4.778	0.000	0.044	0.116	0.044	0.116

Fig. A.18 Answer to Chapter 6: Practice Problem #3

Chapter 6: *Practice Problem #3 (continued)*

1. r = +.78
2. a = y-intercept = −36.277
3. b = slope = + 0.080
4. Y = a + b X
   Y = −36.277 + 0.080 X
5. Y = −36.277 + 0.080 (800)
   Y = −36.277 + 64
   Y = 27.72 degrees centigrade

Chapter 7: *Practice Problem #1 Answer (see Fig. A.19)*

| Research question: | "How well does gunpowder particle size predict the breech pressure in a .50 caliber rifle?" | | | |

Breech pressure (Mpa)	Round (gr)	Cylinder (gr)	Flake (gr)	Irregular (gr)
365	73	53	44	44
375	72	52	44	46
380	75	51	38	50
375	55	75	37	47
370	52	72	41	49
370	53	71	42	48
375	47	48	44	75
365	43	43	51	77
360	51	54	38	71
355	53	50	71	40
360	44	43	73	54
360	45	48	79	42

SUMMARY OUTPUT

Regression Statistics	
Multiple R	0.74
R Square	0.545
Adjusted R Square	0.249
Standard Error	6.199
Observations	12

ANOVA

	df	SS	MS	F	Significance F
Regression	4	367.609	91.902	3.189	0.086
Residual	8	307.391	38.424		
Total	12	675.000			

	Coefficients	Standard Error	t Stat
Intercept	377.0213	29.450	12.802
Round (gr)	0.1347	0.247	0.544
Cylinder (gr)	0.0000	0.000	65535.000
Flake (gr)	-0.3236	0.177	-1.827
Irregular (gr)	-0.0136	0.201	-0.068

	Breech pressure (Mpa)	Round (gr)	Cylinder (gr)	Flake (gr)	Irregular (gr)
Breech pressure (Mpa)	1				
Round (gr)	0.50	1			
Cylinder (gr)	0.37	0.12	1		
Flake (gr)	-0.71	-0.47	-0.54	1	
Irregular (gr)	0.06	-0.44	-0.34	-0.28	1

Fig. A.19 Answer to Chapter 7: Practice Problem #1

Chapter 7: *Practice Problem #1 (continued)*

1. Multiple correlation $= .74$
2. y-intercept $= 377.0213$
3. $b_1 = 0.1347$
4. $b_2 = 0.0000$
5. $b_3 = -0.3236$
6. $b_4 = -0.0136$
7. $Y = a + b_1 X_1 + b_2 X_2 + b_3 X_3 + b_4 X_4$
   $Y = 377.0213 + 0.1347 X_1 + 0.0000 X_2 - 0.3236 X_3 - 0.0136 X_4$
8. $Y = 377.0213 + 0.1347 (63) + 0.0000 (58) - 0.3236 (41) - 0.0136 (50)$
   $Y = 377.0213 + 8.49 + 0.0 - 13.27 - 0.68$
   $Y = 371.56$
   $Y = 372$ Mpa
9. 0.50
10. 0.37
11. $-.71$
12. .06
13. .12
14. $-.54$
15. The best predictor of breech pressure was flake ($r = -.71$). Remember: *You need to ignore the negative sign!*
16. The four predictors combined predict breech pressure at $R_{xy} = .74$, and this is slightly better than the best single predictor by itself.

Chapter 7: *Practice Problem #2 Answer (see Fig. A.20)*

STOPPING DISTANCE (feet)	ENFORCEMENT SPEED (mph)	WEIGHT (tons)
45	2.9	3300
65	2.6	3800
85	4.2	3600
95	7.2	3750
100	8.6	3850
125	4.6	3950
135	12.2	4320
165	13.2	3850
186	6.5	3600
286	11.1	4750
320	15.6	3850
586	20.1	5800
650	26.2	5500

SUMMARY OUTPUT

Regression Statistics	
Multiple R	0.95
R Square	0.910
Adjusted R Square	0.892
Standard Error	63.973
Observations	13

ANOVA

	df	SS	MS	F	Significance F
Regression	2	415594.879	207797.440	50.774	5.79013E-06
Residual	10	40925.890	4092.589		
Total	12	456520.769			

	Coefficients	Standard Error	t Stat	P-value	Lower 95%
Intercept	-412.989	145.361	-2.841	0.018	-736.874
ENFORCEMENT SPEED (mph)	15.305	4.695	3.260	0.009	4.845
WEIGHT (tons)	0.114	0.044	2.594	0.027	0.016

	STOPPING DISTANCE (feet)	ENFORCEMENT SPEED (mph)	WEIGHT (tons)
STOPPING DISTANCE (feet)	1		
ENFORCEMENT SPEED (mph)	0.92	1	
WEIGHT (tons)	0.90	0.83	1

Fig. A.20   Answer to Chapter 7: Practice Problem #2

Chapter 7: *Practice Problem #2 (continued)*

1. $R_{xy} = .95$
2. $a = $ y-intercept $= -412.989$
3. $b_1 = 15.305$
4. $b_2 = 0.114$
5. $Y = a + b_1 X_1 + b_2 X_2$
   $Y = -412.989 + 15.305 X_1 + 0.114 X_2$
6. $Y = -412.989 + 15.305 (10.8) + 0.114 (4600)$
   $Y = -412.989 + 165.294 + 524.4$
   $Y = 276.705$ feet
7. $+ 0.92$
8. $+ 0.90$
9. $+ 0.83$
10. Enforcement speed is the better predictor of stopping distance ($r = +.92$)
11. The two predictors combined predict stopping distance slightly better ($R_{xy} = .95$) than the better single predictor by itself

Chapter 7: *Practice Problem #3 Answer (see Fig. A.21)*

CRYSTAL GROWTH RATES

GROWTH (cm)	Temperature ($^\circ$C)	Humidity (%)	Time (days)
2.50	20	30	4
3.00	25	40	7
3.50	22	32	10
3.00	26	36	4
2.40	25	34	6
4.50	23	40	7
3.50	27	42	9
2.50	29	36	7
4.25	28	40	12
5.00	31	43	10

SUMMARY OUTPUT

Regression Statistics	
Multiple R	0.81
R Square	0.658
Adjusted R Square	0.487
Standard Error	0.652
Observations	10

ANOVA

	df	SS	MS	F	Significance F
Regression	3	4.902	1.634	3.846	0.075
Residual	6	2.549	0.425		
Total	9	7.450			

	Coefficients	Standard Error	t Stat	P-value	Lower 95%	Upper 95%
Intercept	-0.859	1.990	-0.432	0.681	-5.727	4.009
Temperature ($^\circ$C)	-0.084	0.092	-0.910	0.398	-0.311	0.142
Humidity (%)	0.140	0.073	1.911	0.104	-0.039	0.319
Time (days)	0.160	0.097	1.643	0.152	-0.078	0.399

	GROWTH (cm)	Temperature ($^\circ$C)	Humidity (%)	Time (days)
GROWTH (cm)	1			
Temperature ($^\circ$C)	0.38	1		
Humidity (%)	0.69	0.70	1	
Time (days)	0.67	0.46	0.51	1

Fig. A.21 Answer to Chapter 7: Practice Problem #3

Chapter 7: *Practice Problem #3 (continued)*

1. Multiple correlation = .81
2. a = y-intercept = −0.859
3. $b_1 = -0.084$
4. $b_2 = 0.140$
5. $b_3 = 0.160$
6. $Y = a + b_1 X_1 + b_2 X_2 + b_3 X_3$
   $Y = -0.859 - 0.084 X_1 + 0.140 X_2 + 0.160 X_3$
7. $Y = -0.859 - 0.084 (25) + 0.140 (34) + 0.160 (6)$
   $Y = -0.859 - 2.1 + 4.76 + 0.96$
   $Y = 2.76$ cm
8. + 0.38
9. + 0.69
10. + 0.67
11. + 0.70
12. + 0.46
13. + 0.51
14. The best single predictor of GROWTH was Humidity (r = .69).
15. The three predictors combined predict GROWTH at $R_{xy} = .81$, and this is much better than the best single predictor by itself.

Chapter 8: *Practice Problem #1 Answer (see Fig. A.22)*

CHEMICAL REACTION YIELD (grams of product produced)

BELOW ROOM TEMP (15 °C)	ROOM TEMP (25 °C)	ABOVE ROOM TEMP (30 °C)
90	85	76
85	89	80
74	83	90
89	79	84
84	74	78
95	75	65
92	86	42
65	87	58
75	86	63
73	88	75
54		66
71		

Anova: Single Factor

SUMMARY

Groups	Count	Sum	Average	Variance
BELOW ROOM TEMP (15 °C)	12	947	78.92	151.72
ROOM TEMP (25 °C)	10	832	83.20	28.84
ABOVE ROOM TEMP (30 °C)	11	777	70.64	183.45

ANOVA

Source of Variation	SS	df	MS	F	P-value	F crit
Between Groups	867.12	2	433.56	3.46	0.04	3.32
Within Groups	3763.06	30	125.44			
Total	4630.18	32				

ROOM TEMP vs. ABOVE ROOM TEMP

1/ n ROOM TEMP + 1/ n ABOVE ROOM TEMP	0.19
s. e. ROOM TEMP vs. ABOVE ROOM TEMP	4.89
ANOVA t-test	2.57

Fig. A.22 Answer to Chapter 8: Practice Problem #1

Chapter 8: *Practice Problem #1 (continued)*

Let Group 1 = BELOW ROOM TEMP, Group 2 = ROOM TEMP, and Group 3 = ABOVE ROOM TEMP

1. $H_0 : \mu_1 = \mu_2 = \mu_3$
   $H_1 : \mu_1 \neq \mu_2 \neq \mu_3$
2. $MS_b = 433.56$
3. $MSw = 125.44$
4. $F = 433.56 / 125.44 = 3.46$
5. critical $F = 3.32$
6. Result: Since 3.46 is greater than 3.32, we reject the null hypothesis and accept the research hypothesis
7. There was a significant difference between the three temperatures in the grams of product produced.

ROOM TEMP vs. ABOVE ROOM TEMP

8. $H_0 : \mu_2 = \mu_3$
   $H_1 : \mu_2 \neq \mu_3$
9. 83.20
10. 70.64
11. df $= 33 - 3 = 30$
12. critical t $= 2.042$
13. $1/10 + 1/11 = 0.10 + 0.09 = 0.19$
    s.e. $=$ SQRT $(125.44 * 0.19) =$ SQRT $(23.83) = 4.88$
14. ANOVA t $= (83.20 - 70.64) / 4.88 = 2.57$
15. Result: Since the absolute value of 2.57 is greater than 2.042, we reject the null hypothesis and accept the research hypothesis
16. Conclusion: ROOM TEMP produced significantly more grams of product than ABOVE ROOM TEMP (83.2 vs. 70.6).

Chapter 8: *Practice Problem #2 Answer (see Fig. A.23)*

HORSEPOWER OUTPUT WITH PROTOTYPE FUEL INJECTOR

FUEL INJECTOR A	FUEL INJECTOR B	FUEL INJECTOR C	FUEL INJECTOR D
550	550	555	555
600	560	565	570
650	580	575	580
625	600	580	575
620	610	590	585
670	630	625	590
680	660	640	610
700	670	665	630
725	690	680	625
710		690	645
		695	690
			680

Anova: Single Factor

SUMMARY

Groups	Count	Sum	Average	Variance
FUEL INJECTOR A	10	6530	653.00	2995.56
FUEL INJECTOR B	9	5550	616.67	2450.00
FUEL INJECTOR C	11	6860	623.64	2820.45
FUEL INJECTOR D	12	7335	611.25	1900.57

ANOVA

Source of Variation	SS	df	MS	F	P-value	F crit
Between Groups	10739.32	3	3579.77	1.42	0.25	2.85
Within Groups	95670.80	38	2517.65			
Total	106410.12	41				

IMPORTANT NOTE: Since the value of F of 1.42 is less than the critical F of 2.85, you cannot run ANY of the paired comparisons for the Fuel Injectors since you are accepting the null hypothesis than there is no difference in horsepower output between the four types of fuel injectors.

Fig. A.23 Answer to Chapter 8: Practice Problem #2

Chapter 8: *Practice Problem #2 (continued)*

1. Null hypothesis: $\mu_A = \mu_B = \mu_C = \mu_D$
   Research hypothesis: $\mu_A \neq \mu_B \neq \mu_C \neq \mu_D$
2. $MS_b = 3579.77$
3. $MS_w = 2517.65$
4. $F = 3579.77 / 2517.65 = 1.42$
5. Critical $F = 2.85$
6. Since the F-value of 1.42 is less than the critical F value of 2.85, we accept the null hypothesis.
7. There was no difference between the four types of fuel injectors in their horsepower output.
8. 8 – 16. *Be careful here!* You need to remember that it is incorrect to perform ANY ANOVA t-test when the value of F is less than the critical value of F. The ANOVA F-test found no difference between the four types of fuel injectors in horsepower output, and, therefore, *you cannot compare any two injectors using the ANOVA t-test!*

Chapter 8: *Practice Problem #3 Answer (see Fig. A.24)*

HIGHWAY MILES PER GALLON (mpg) COMPARISON OF FIVE TYPES OF CARS

	1	2	3	4	5
	SUBCOMPACTS (mpg)	COMPACTS (mpg)	MID-SIZE (mpg)	LARGE (mpg)	SUVs (mpg)
	28.1	26.2	24.0	22.0	18.1
	30.2	28.3	26.3	23.1	20.2
	29.3	29.3	25.2	25.4	22.3
	31.6	27.0	27.1	24.3	21.4
	33.0	28.0	28.0	25.0	20.5
	34.3	29.5	23.6	24.7	19.0
	32.1	31.0	29.2	23.1	18.2
	35.0	32.3		22.4	19.1
		33.1		26.0	
				21.3	

Anova: Single Factor

SUMMARY

Groups	Count	Sum	Average	Variance
SUBCOMPACTS	8	253.60	31.70	5.78
COMPACTS	9	264.70	29.41	5.48
MID-SIZE	7	183.40	26.20	4.28
LARGE	10	237.30	23.73	2.48
SUVs	8	158.80	19.85	2.29

ANOVA

Source of Variation	SS	df	MS	F	P-value	F crit
Between Groups	718.23	4	179.56	44.80	1.07093E-13	2.63
Within Groups	148.29	37	4.01			
Total	866.52	41				

COMPACTS vs. LARGE

1/9+1/10	0.21
s.e. ANOVA	0.92
ANOVA t-test	6.18

Fig. A.24   Answer to Chapter 8: Practice Problem #3

Chapter 8: *Practice Problem #3 (continued)*

Let SUBCOMPACTS = Group 1, COMPACTS = Group 2, MID-SIZE = Group 3, LARGE = Group 4, and SUVs = Group 5

1. Null hypothesis: $\mu_1 = \mu_2 = \mu_3 = \mu_4 = \mu_5$
   Research hypothesis: $\mu_1 \neq \mu_2 \neq \mu_3 \neq \mu_4 \neq \mu_5$
2. $MS_b = 179.56$
3. $MS_w = 4.01$
4. $F = 179.56 / 4.01 = 44.78$
5. critical $F = 2.63$
6. Result: Since the F-value of 44.78 is greater than the critical F value of 2.63, we reject the null hypothesis and accept the research hypothesis.
7. Conclusion: There was a significant difference between the five types of vehicles in their highway miles per gallon.
8. Null hypothesis: $\mu_2 = \mu_4$
   Research hypothesis: $\mu_2 \neq \mu_4$
9. 29.41
10. 23.73
11. degrees of freedom $= 42 - 5 = 37$
12. critical $t = 2.026$
13. $s.e._{ANOVA} = SQRT(MS_w \times \{1/9 + 1/10\}) = SQRT (4.01 \times 0.21) = SQRT (0.84) = 0.92$
14. ANOVA $t = (29.41 - 23.73) / .92 = 6.17$
15. Since the absolute value of 6.17 is greater than the critical t of 2.026, we reject the null hypothesis and accept the research hypothesis.
16. COMPACTS had significantly higher highway mpg than LARGE vehicles (29.4 vs. 23.7)

Appendix B
Practice Test

Chapter 1: *Practice Test*

Suppose that you were hired as a research assistant on a project involving concrete blocks, and that your responsibility on this team was to measure the compressive strength in units of 100 pounds per square inch (psi) of concrete blocks from a certain supplier. You want to try out your Excel skills on a small random sample of blocks. The hypothetical data is given below (see Fig. B.1).

COMPRESSIVE STRENGTH OF CONCRETE BLOCKS

100 POUNDS PER SQUARE INCH (psi)
39.3
42.6
54.7
51.3
48.4
46.8
39.8
40.5
42.5
50.6
51.4
53.7
48.5

Fig. B.1 Worksheet Data for Chapter 1 Practice Test (Practical Example)

T.J. Quirk et al., *Excel 2010 for Physical Sciences Statistics: A Guide to Solving Practical Problems*, DOI 10.1007/978-3-319-00630-7,
© Springer International Publishing Switzerland 2013

(a) Create an Excel table for these data, and then use Excel to the right of the table to find the sample size, mean, standard deviation, and standard error of the mean for these data. Label your answers, and round off the mean, standard deviation, and standard error of the mean to two decimal places.

(b) Save the file as: CONCRETE3

Chapter 2: *Practice Test*

Suppose that an engineer who works for an automobile manufacturer wants to take a random sample of 12 of the 54 engine crankshaft bearings produced during the last shift in the plant to see how many of them had a surface finish that was rougher than the engineering specifications required.

(a) Set up a spreadsheet of frame numbers for these bearings with the heading: FRAME NUMBERS

(b) Then, create a separate column to the right of these frame numbers which duplicates these frame numbers with the title: Duplicate frame numbers.

(c) Then, create a separate column to the right of these duplicate frame numbers called RAND NO. and use the =RAND() function to assign random numbers to all of the frame numbers in the duplicate frame numbers column, and change this column format so that 3 decimal places appear for each random number.

(d) Sort the *duplicate frame numbers and random numbers* into a random order.

(e) Print the result so that the spreadsheet fits onto one page.

(f) Circle on your printout the I.D. number of the first 12 engine crankshaft bearings that you would use in your test.

(g) Save the file as: RAND62

> *Important note:* Note that everyone who does this problem will generate a different random order of bearings ID numbers since Excel assign a different random number each time the RAND() command is used. For this reason, the answer to this problem given in this Excel Guide will have a completely different sequence of random numbers from the random sequence that you generate. This is normal and what is to be expected.

Chapter 3: *Practice Test*

Suppose that a manufacturer of a certain type of house paint has a factory that produced an average of 60 tons per day over the past month for this paint. Suppose, further, that this factory tries out a new manufacturing process for this type of paint for 30 days. You have been asked to "run the data" to see if any change has occurred in the production output with this new procedure, and you have decided to test your Excel skills on a random sample of hypothetical data given in Fig. B.2

Fig. B.2 Worksheet Data
for Chapter 3 Practice Test
(Practical Example)

HOUSE PAINT PRODUCTION

New manufacturing process (tons/day)
62
61
64
63
61
63
62
67
59
61
60
63
64
62
65

(a) Create an Excel table for these data, and use Excel to the right of the table to find the sample size, mean, standard deviation, and standard error of the mean for these data. Label your answers, and round off the mean, standard deviation, and standard error of the mean to two decimal places in number format.
(b) By hand, write the null hypothesis and the research hypothesis on your printout.
(c) Use Excel's *TINV function* to find the 95% confidence interval about the mean for these data. Label your answers. Use two decimal places for the confidence interval figures in number format.
(d) On your printout, draw a diagram of this 95% confidence interval by hand, including the reference value.

(e) On your spreadsheet, enter the *result*.

(f) On your spreadsheet, enter the *conclusion in plain English*.

(g) Print the data and the results so that your spreadsheet fits onto one page.

(h) Save the file as: PAINT15

Chapter 4: *Practice Test*

Suppose that you work for a company that manufactures small submersible pumps. Submersible pumps are pumps that can be submerged under water and they are used to pump water out of an area. For example, submersible pumps can be used to pump flood water out of basements. Suppose, further, that your company has developed a new style of pump and has decided to test it on some recently flooded homes near Grafton, Illinois, in the USA. The old style pumps pumped an average of 46 gallons per minute (gal/min). You want to test your Excel skills on a small sample of data using the hypothetical data given in Fig. B.3.

Fig. B.3 Worksheet Data for Chapter 4 Practice Test (Practical Example)

OUTPUT OF NEW SUBMERSIBLE PUMP
GALLONS PER MINUTE (gal/min)
51
50
50
49
50
48
52
50
50
49
48
49
50
51
49
50
49
51
51
50

(a) Write the null hypothesis and the research hypothesis on your spreadsheet.
(b) Create a spreadsheet for these data, and then use Excel to find the sample size, mean, standard deviation, and standard error of the mean to the right of the data set. Use number format (2 decimal places) for the mean, standard deviation, and standard error of the mean.
(c) Type the *critical t* from the t-table in Appendix E onto your spreadsheet, and label it.
(d) Use Excel to compute the t-test value for these data (use 2 decimal places) and label it on your spreadsheet.
(e) Type the *result* on your spreadsheet, and then type the *conclusion in plain English* on your spreadsheet.
(f) Save the file as: PUMP8

Chapter 5: *Practice Test*

Suppose that an automobile repair parts manufacturer/supplier wants to test the crash resistance of two brands of front-bumpers for 2-door passenger sedans (BRAND X and BRAND Y). The engineer in charge of this project has decided to test these bumpers on 2013 Honda Civics that are purposely crashed into a cement wall at a speed of 15 miles per hour (mph), and then to estimate the cost of repairs to the front bumper after this test. The engineer then wants to test her Excel skills on the hypothetical data given in Fig. B.4.

FRONT-BUMPER CRASH RESISTANT TEST (2-door passenger car)	

2013 Honda Civic: 15 mph speed

REPAIR ESTIMATE ($)

BRAND X	BRAND Y
1,242	1,312
1,264	1,300
1,231	1,295
1,159	1,395
1,015	1,354
1,135	1,368
1,140	1,412
1,253	1,295
1,264	1,275
1,275	1,300
1,283	1,354

Fig. B.4 Worksheet Data for Chapter 5 Practice Test (Practical Example)

(a) Write the null hypothesis and the research hypothesis.
(b) Create an Excel table that summarizes these data.
(c) Use Excel to find the standard error of the difference of the means.
(d) Use Excel to perform a *two-group t-test*. What is the value of *t* that you obtain (use two decimal places)?
(e) On your spreadsheet, type the *critical value of t* using the t-table in Appendix E.
(f) Type the *result* of the test on your spreadsheet.
(g) Type your *conclusion in plain English* on your spreadsheet.
(h) Save the file as: BUMPER3
(i) Print the final spreadsheet so that it fits onto one page.

Chapter 6: *Practice Test*

What is the relationship between the weight of the car (measured in thousands of pounds) and its city miles per gallon (mpg) in 4-door passenger sedans? Suppose that you wanted to study this question using different models of cars. Analyze the hypothetical data that are given in Fig. B.5.

Research question: "What is the relationship between the weight of a 4-door sedan and its miles per gallon (mpg) performance in city driving?"

Weight (1000 lbs)	City Miles Per Gallon (mpg)
2.1	32.2
2.4	28.6
3.5	26.7
2.3	28.1
3.4	27.7
4.1	16.2
3.8	20.9
3.6	22.4
4.3	18.4
4.2	15.3

Fig. B.5 Worksheet Data for Chapter 6 Practice Test (Practical Example)

Create an Excel spreadsheet, and enter the data.

(a) create an *XY scatterplot* of these two sets of data such that:

- top title: RELATIONSHIP BETWEEN WEIGHT AND CITY mpg IN 4-DOOR SEDANS
- x-axis title: WEIGHT (1000 lbs)
- y-axis title: CITY MILES PER GALLON (mpg)
- move the chart below the table
- re-size the chart so that it is 7 columns wide and 25 rows long
- delete the legend
- delete the gridlines

(b) Create the *least-squares regression line* for these data on the scatterplot.
(c) Use Excel to run the regression statistics to find the *equation for the least-squares regression line* for these data and display the results below the chart on your spreadsheet. Add the regression equation to the chart. Use number format (3 decimal places) for the correlation and for the coefficients

Print *just the input data and the chart* so that this information fits onto one page in portrait format.

Then, print *just the regression output table* on a separate page so that it fits onto that separate page in portrait format.

By hand:

(d) Circle and label the value of the *y-intercept* and the *slope* of the regression line on your printout.
(e) Write the regression equation *by hand* on your printout for these data (use three decimal places for the y-intercept and the slope).

(f) Circle and label the *correlation* between the two sets of scores in the regression analysis summary output table on your printout.
(g) Underneath the regression equation you wrote by hand on your printout, use the regression equation to predict the average city mpg of a 4-door sedan that weighted 2,500 pounds.
(h) *Read from the graph,* the average city mpg you would predict for a 4-door sedan that weighed 3,600 pounds, and write your answer in the space immediately below:_____
(i) save the file as: sedan3

Chapter 7: *Practice Test*

Suppose that you wanted to estimate the total number of gallons required for 2013 4-door sedans when they were driven on a specific route of 200 miles between St. Louis, Missouri, and Indianapolis, Indiana, at specified speeds using drivers that were about the same weight. You have decided to use two predictors: (1) weight of the car (measured in thousands of pounds), and (2) the car's engine horsepower. To check your skills in Excel, you have created the hypothetical data given in Fig. B.6.

TOTAL GALLONS USED TO DRIVE FROM ST. LOUIS TO INDIANAPOLIS

2013 FOUR-DOOR SEDANS

TOTAL GALLONS USED	WEIGHT (1000 lbs)	HORSEPOWER
6.1	3.8	130
6.3	3.7	150
4.8	4.0	140
4.2	2.4	125
3.8	2.9	98
4.7	3.0	115
3.5	2.1	121
5.5	2.9	123
5.9	3.1	110
3.4	2.1	96

Fig. B.6 Worksheet Data for Chapter 7 Practice Test (Practical Example)

(a) create an Excel spreadsheet using TOTAL GALLONS USED as the criterion (Y), and the other variables as the two predictors of this criterion ($X_1 =$ WEIGHT (1000 lbs), and $X_2 =$ HORSEPOWER).
(b) Use Excel's *multiple regression* function to find the relationship between these three variables and place the SUMMARY OUTPUT below the table.

(c) Use number format (2 decimal places) for the multiple correlation on the Summary Output, and use two decimal places for the coefficients in the SUMMARY OUTPUT.

(d) Save the file as: GALLONS9

(e) Print the table and regression results below the table so that they fit onto one page.

Answer the following questions using your Excel printout:

1. What is the multiple correlation R_{xy} ?
2. What is the y-intercept a ?
3. What is the coefficient for WEIGHT b_1 ?
4. What is the coefficient for HORSEPOWER b_2 ?
5. What is the multiple regression equation?
6. Predict the TOTAL GALLONS USED you would expect for a WEIGHT of 3,800 pounds and a car that had 126 HORSEPOWER.

(f) Now, go back to your Excel file and create a correlation matrix for these three variables, and place it underneath the SUMMARY OUTPUT.

(g) Re-save this file as: GALLONS9

(h) Now, print out *just this correlation matrix* on a separate sheet of paper.

Answer to the following questions using your Excel printout. (Be sure to include the plus or minus sign for each correlation):

7. What is the correlation between WEIGHT and TOTAL GALLONS USED?
8. What is the correlation between HORSEPOWER and TOTAL GALLONS USED?
9. What is the correlation between WEIGHT and HORSEPOWER?
10. Discuss which of the two predictors is the better predictor of total gallons used.
11. Explain in words how much better the two predictor variables combined predict total gallons used than the better single predictor by itself.

Chapter 8: *Practice Test*

Let's consider an experiment in which you want to compare the strength of beams made of three types of materials: (1) steel, (2) Alloy A, and (3) Alloy B. The strength of the material was measured by placing each beam in a horizontal position with a support on each end, and then applying a force of 2,500 pounds to the center of each beam. The "deflection of the beam" was then measured in $1/1000^{th}$ of an inch. You decide to test your Excel skills on a small sample of beams, and you have created the hypothetical data given in Fig. B.7.

STEEL	ALLOY A	ALLOY B
81	76	78
85	78	79
86	79	81
84	81	80
87	82	83
82	84	82
81	83	81
85	78	79
86	79	78
88	81	80
87	78	81
86	77	
	76	

STRENGTH OF THREE TYPES OF BEAMS

(measured in $1/1000^{th}$ of an inch in deflection)

Fig. B.7 Worksheet Data for Chapter 8 Practice Test (Practical Example)

(a) Enter these data on an Excel spreadsheet.

Let STEEL = Group 1, ALLOY A = Group 2, and ALLOY B = Group 3.

(b) On your spreadsheet, write the null hypothesis and the research hypothesis for these data

(c) Perform a *one-way ANOVA test* on these data, and show the resulting ANOVA table underneath the input data for the three types of beams.

(d) If the F-value in the ANOVA table is significant, create an Excel formula to compute the ANOVA t-test comparing the STEEL beams versus the ALLOY A beams, and show the results below the ANOVA table on the spreadsheet (put the standard error and the ANOVA t-test value on separate lines of your spreadsheet, and use two decimal places for each value)

(e) Print out the resulting spreadsheet so that all of the information fits onto one page

(f) On your printout, label by hand the MS (between groups) and the MS (within groups)

(g) Circle and label the value for F on your printout for the ANOVA of the input data

(h) Label by hand on the printout the mean for steel beams and the mean for Alloy
A beams that were produced by your ANOVA formulas

(i) Save the spreadsheet as: STRENGTH3

On a separate sheet of paper, now do the following by hand:

(j) find the critical value of F in the ANOVA Single Factor results table

(k) write a summary of the *result* of the ANOVA test for the input data

(l) write a summary of the *conclusion* of the ANOVA test in plain English for the
input data

(m) write the null hypothesis and the research hypothesis comparing steel beams
versus Alloy A beams.

(n) compute the degrees of freedom for the *ANOVA t-test* by hand for three types
of beams.

(o) use your calculator and Excel to compute the standard error (s.e.) of the
ANOVA t-test

(p) Use your calculator and Excel to compute the ANOVA t-test value

(q) write the *critical value of t* for the ANOVA t-test using the table in Appendix E.

(r) write a summary of the *result* of the ANOVA t-test

(s) write a summary of the *conclusion* of the ANOVA t-test in plain English

Appendix C
Answers to Practice Test

Practice Test Answer: *Chapter 1 (see. Fig. C.1)*

COMPRESSIVE STRENGTH OF CONCRETE BLOCKS		
100 POUNDS PER SQUARE INCH (psi)		
39.3		
42.6	n	13
54.7		
51.3		
48.4	Mean	46.93
46.8		
39.8		
40.5	STDEV	5.42
42.5		
50.6		
51.4	s.e.	1.50
53.7		
48.5		

Fig. C.1 Practice Test Answer to Chapter 1 Problem

T.J. Quirk et al., *Excel 2010 for Physical Sciences Statistics: A Guide to Solving Practical Problems*, DOI 10.1007/978-3-319-00630-7,
© Springer International Publishing Switzerland 2013

Practice Test Answer: *Chapter 2 (see. Fig. C.2)*

Fig. C.2 Practice Test
Answer to Chapter 2 Problem

FRAME NUMBERS	Duplicate frame numbers	RAND NO.
1	5	0.679
2	11	0.926
3	40	0.965
4	27	0.556
5	54	0.022
6	1	0.062
7	29	0.140
8	33	0.559
9	31	0.646
10	17	0.942
11	51	0.822
12	12	0.537
13	24	0.179
14	45	0.543
15	7	0.548
16	32	0.936
17	36	0.831
18	38	0.804
19	8	0.667
20	52	0.299
21	18	0.992
22	13	0.910
23	28	0.352
24	14	0.055
25	25	0.730
26	34	0.565
27	35	0.261
28	53	0.234
29	26	0.878
30	44	0.915
31	15	0.058
32	50	0.523
33	46	0.673
34	48	0.341
35	21	0.898
36	39	0.618
37	16	0.891
38	37	0.121
39	43	0.563
40	20	0.027
41	30	0.307
42	49	0.751
43	6	0.853
44	47	0.044
45	2	0.534
46	10	0.399
47	42	0.914
48	23	0.120
49	22	0.071
50	41	0.772
51	3	0.957
52	9	0.252
53	19	0.301
54	4	0.186

Practice Test Answer: *Chapter 3 (see. Fig. C.3)*

HOUSE PAINT PRODUCTION

New manufacturing process (tons/day)				
62				
61	Null hypothesis:	μ	=	60 tons/day
64				
63	Research hypothesis:	μ	$\neq$	60 tons/day
61				
63	n	15		
62				
67	Mean	62.47		
59				
61	STDEV	2.03		
60				
63	s.e.	0.52		
64				
62	95% CONFIDENCE INTERVAL			
65				
	Lower limit		61.34	
	Upper limit		63.59	

```
---------60 ---------------- 61.34 ----- --------------- 62.47 ----- ---------------- 63.59 ----
         Ref.                lower                        Mean                      upper
         value               limit                                                 limit
```

Result: Since the reference value is outside the confidence interval, we reject the null hypothesis and accept the research hypothesis.

Conclusion: The new manufacturing process produced significantly more than 60 tons/day, and it was probably closer to 62.5 tons/day.

Fig. C.3 Practice Test Answer to Chapter 3 Problem

Practice Test Answer: *Chapter 4 (see. Fig. C.4)*

OUTPUT OF NEW SUBMERSIBLE PUMP				
GALLONS PER MINUTE (gal/min)				
51				
50				
50	Null hypothesis:	μ	=	46 gal/min
49				
50	Research hypothesis:	μ	≠	46 gal/min
48				
52				
50	n	20		
50				
49				
48	Mean	49.85		
49				
50				
51	STDEV	1.04		
49				
50				
49	s.e.	0.23		
51				
51				
50	critical t	2.093		
	t-test	16.56		

Result: Since the absolute value of 16.56 is greater than the critical t of 2.093,
we reject the null hypothesis and accept the research hypothesis.

Conclusion: The new style of submersible pumps pumped significantly more water
than the older style of pumps, and it was probably closer to almost 50 gal/min.

Fig. C.4 Practice Test Answer to Chapter 4 Problem

Practice Test Answer: *Chapter 5 (see. Fig. C.5)*

FRONT-BUMPER CRASH RESISTANT TEST (2-door passenger car)

2013 Honda Civic: 15 mph speed

Null hypothesis:	μ_1	$=$		μ_2
Research hypothesis:	μ_1	$\neq$		μ_2

REPAIR ESTIMATE ($)

BRAND X	BRAND Y
1,242	1,312
1,264	1,300
1,231	1,295
1,159	1,395
1,015	1,354
1,135	1,368
1,140	1,412
1,253	1,295
1,264	1,275
1,275	1,300
1,283	1,354

BRAND	n	Mean ($)	STDEV ($)
1 BRAND X	11	1,205.55	83.45
2 BRAND Y	11	1,332.73	45.89

(n1 - 1) x STDEV1 squared	69,632.73
(n2 - 1) x STDEV2 squared	21,062.18
n1 + n2 - 2	20
1/n1 + 1/n2	0.18
s.e.	28.71
critical t	2.086
t-test	-4.43

Result: Since the absolute value of − 4.43 is greater than the critical t of 2.086, we reject the null hypothesis and accept the research hypothesis.

Conclusion: For the 2013 Honda Civic (2-door) sedan, the front-bumper for Brand X cost significantly less to repair than the front-bumper of Brand Y ($1,206 vs. $1,333).

Fig. C.5 Practice Test Answer to Chapter 5 Problem

Practice Test Answer: *Chapter 6 (see. Fig. C.6)*

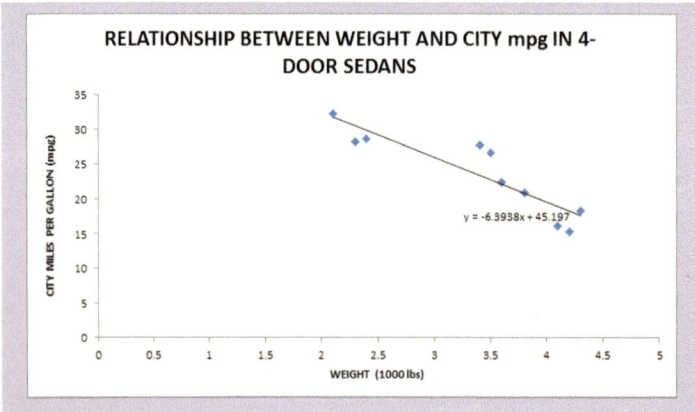

| Research question: | "What is the relationship between the weight of a 4-door sedan and its miles per gallon (mpg) performance in city driving?" |

Weight (1000 lbs)	City Miles Per Gallon (mpg)
2.1	32.2
2.4	28.6
3.5	26.7
2.3	28.1
3.4	27.7
4.1	16.2
3.8	20.9
3.6	22.4
4.3	18.4
4.2	15.3

RELATIONSHIP BETWEEN WEIGHT AND CITY mpg IN 4-DOOR SEDANS

$y = -6.3938x + 45.197$

SUMMARY OUTPUT

Regression Statistics	
Multiple R	0.900
R Square	0.8
Adjusted R Square	0.8
Standard Error	2.7
Observations	10

ANOVA

	df	SS	MS	F	Significance F
Regression	1	247.0	247.0	34.0	0.0004
Residual	8	58.1	7.3		
Total	9	305.0			

	Coefficients	Standard Error	t Stat	P-value	Lower 95%	Upper 95%
Intercept	45.197	3.8	11.9	0.0	36.5	53.9
X Variable 1	-6.394	1.1	-5.8	0.0	-8.9	-3.9

Fig. C.6 Practice Test Answer to Chapter 6 Problem

Practice Test Answer: *Chapter 6: (continued)*

(d) a = y-intercept = 45.197
     b = slope = −6.394 (note the negative sign!)
(e) Y = a + b X
     Y = 45.197 −6.394 X
(f) r = correlation = −.900 (note the negative sign!)
(g) Y = 45.197 −6.394 (2.5)
     Y = 45.197 −15.985
     Y = 29.212 mpg
(h) About 22 – 23 mpg

Practice Test Answer: *Chapter 7 (see. Fig. C.7)*

TOTAL GALLONS USED TO DRIVE FROM ST. LOUIS TO INDIANAPOLIS

2013 FOUR-DOOR SEDANS

TOTAL GALLONS USED	WEIGHT (1000 lbs)	HORSEPOWER
6.1	3.8	130
6.3	3.7	150
4.8	4.0	140
4.2	2.4	125
3.8	2.9	98
4.7	3.0	115
3.5	2.1	121
5.5	2.9	123
5.9	3.1	110
3.4	2.1	96

SUMMARY OUTPUT

Regression Statistics	
Multiple R	0.77
R Square	0.593
Adjusted R Square	0.477
Standard Error	0.787
Observations	10

ANOVA

	df	SS	MS	F	Significance F
Regression	2	6.320	3.160	5.102	0.043
Residual	7	4.336	0.619		
Total	9	10.656			

	Coefficients	Standard Error	t Stat	P-value	Lower 95%
Intercept	0.29	1.877	0.154	0.882	-4.150
WEIGHT (1000 lbs)	1.01	0.509	1.984	0.088	-0.194
HORSEPOWER	0.01	0.020	0.614	0.559	-0.035

	TOTAL GALLONS USED	WEIGHT (1000 lbs)	HORSEPOWER
TOTAL GALLONS USED	1		
WEIGHT (1000 lbs)	0.76	1	
HORSEPOWER	0.60	0.65	1

Fig. C.7 Practice Test Answer to Chapter 7 Problem

Practice Test Answer: *Chapter 7 (continued)*

1. $R_{xy} = .77$
2. $a = $ y-intercept $= 0.29$
3. $b_1 = 1.01$
4. $b_2 = 0.01$
5. $Y = a + b_1 X_1 + b_2 X_2$
   $Y = 0.29 + 1.01 X_1 + 0.01 X_2$
6. $Y = 0.29 + 1.01 (3.8) + 0.01 (126))$
   $Y = 0.29 + 3.84 + 1.26$
   $Y = 5.39$ gallons
7. $+.76$
8. $+.60$
9. $+.65$
10. The better predictor of TOTAL GALLONS USED was WEIGHT with a correlation of $+.76$.
11. The two predictors combined predict TOTAL GALLONS USED only slightly better ($R_{xy} = .77$) than the better single predictor by itself

Practice Test Answer: *Chapter 8 (see. Fig. C.8)*

Null hypothesis:	μ_1	$=$	μ_2	$=$	μ_3
Research hypothesis:	μ_1	$\neq$	μ_2	$\neq$	μ_3

STRENGTH OF THREE TYPES OF BEAMS

(measured in $1/1000^{th}$ of an inch in deflection)

STEEL	ALLOY A	ALLOY B
81	76	78
85	78	79
86	79	81
84	81	80
87	82	83
82	84	82
81	83	81
85	78	79
86	79	78
88	81	80
87	78	81
86	77	
	76	

Anova: Single Factor

SUMMARY

Groups	Count	Sum	Average	Variance
STEEL	12	1018	84.83	5.61
ALLOY A	13	1032	79.38	6.76
ALLOY B	11	882	80.18	2.56

ANOVA

Source of Variation	SS	df	MS	F	P-value	F crit
Between Groups	210.51	2	105.25	20.63	1.54E-06	3.28
Within Groups	168.38	33	5.10			
Total	378.89	35				

STEEL vs. ALLOY A

1/n STEEL + 1/n ALLOY A	0.16

s.e. STEEL vs. ALLOY A	0.90

ANOVA t-test	6.03

Fig. C.8 Practice Test Answer to Chapter 8 Problem

Practice Test Answer: *Chapter 8 (continued)*

Let STEEL = Group 1, ALLOY A = Group 2, and ALLOY B = Group 3.

(b) $H_0 : \mu_1 = \mu_2 = \mu_3$
$H_1 : \mu_1 \neq \mu_2 \neq \mu_3$

(f) $MS_b = 105.25$ and $MS_w = 5.10$

(g) $F = 20.63$

(h) Mean of STEEL = 84.83 and Mean of ALLOY A = 79.38

(j) critical F = 3.28

(k) Result: Since 20.63 is greater than 3.28, we reject the null hypothesis and accept the research hypothesis

(l) Conclusion: There was a significant difference in strength between the three types of beams.

(m) $H_0 : \mu_1 = \mu_2$
$H_1 : \mu_1 \neq \mu_2$

(n) df = $n_{TOTAL} - k = 36 - 3 = 33$

(o) $1/12 + 1/13 = 0.08 + 0.08 = 0.16$
s.e = SQRT (5.10 * 0.16)
s.e. = SQRT (0.82)
s.e. = 0.90

(p) ANOVA t = (84.83 − 79.38) / 0.90 = 6.06

(q) critical t = 2.035

(r) Result: Since the absolute value of 6.06 is greater than the critical t of 2.035, we reject the null hypothesis and accept the research hypothesis

(s) Conclusion: ALLOY A had significantly less deflection (i.e., it was stronger) than STEEL (79.4 vs. 84.8)

Appendix D
Statistical Formulas

Mean $\qquad\qquad\qquad\qquad\qquad \bar{X} = \dfrac{\sum X}{n}$

Standard Deviation $\qquad\qquad\quad \text{STDEV} = S = \sqrt{\dfrac{\sum (X-\bar{X})^2}{n-1}}$

Standard error of the mean $\qquad s.e. = S_{\bar{X}} = \dfrac{S}{\sqrt{n}}$

Confidence interval about the mean $\quad \bar{X} \pm t\, S_{\bar{X}}$

$$\text{where } S_{\bar{X}} = \dfrac{S}{\sqrt{n}}$$

One-group t-test $\qquad\qquad\qquad t = \dfrac{\bar{X} - \mu}{S_{\bar{X}}}$

$$\text{where } S_{\bar{X}} = \dfrac{S}{\sqrt{n}}$$

Two-group t-test

(a) when both groups have a sample size greater than 30

$$t = \dfrac{\bar{X}_1 - \bar{X}_2}{S_{\bar{X}_1 - \bar{X}_2}}$$

$$\text{where } S_{\bar{X}_1 - \bar{X}_2} = \sqrt{\dfrac{S_1^2}{n_1} + \dfrac{S_2^2}{n_2}}$$

$$\text{and where } df = n_1 + n_2 - 2$$

T.J. Quirk et al., *Excel 2010 for Physical Sciences Statistics: A Guide to Solving Practical Problems*, DOI 10.1007/978-3-319-00630-7,
© Springer International Publishing Switzerland 2013

(b) when one or both groups have a sample size less than 30

$$t = \frac{\bar{X}_1 - \bar{X}_2}{S_{\bar{X}_1 - \bar{X}_2}}$$

where $S_{\bar{X}_1 - \bar{X}_2} = \sqrt{\dfrac{(n_1 - 1)S_1^2 + (n_2 - 1)S_2^2}{n_1 + n_2 - 2}\left(\dfrac{1}{n_1} + \dfrac{1}{n_2}\right)}$

and where $df = n_1 + n_2 - 2$

Correlation $r = \dfrac{\dfrac{1}{n-1}\sum(X - \bar{X})(Y - \bar{Y})}{S_x S_y}$

where S_x = standard deviation of X
and where S_y = standard deviation of Y

Simple linear regression $Y = a + b X$
where a = y-intercept and b = slope of the line

Multiple regression equation $Y = a + b_1 X_1 + b_2 X_2 + b_3 X_3 + $ etc.

where a = y-intercept

One-way ANOVA F-test $F = MS_b / MS_w$

ANOVA t-test $ANOVA\, t = \dfrac{\bar{X}_1 - \bar{X}_2}{s.e._{ANOVA}}$

where $s.e._{ANOVA} = \sqrt{MS_W \left(\dfrac{1}{n_1} + \dfrac{1}{n_2}\right)}$

and where $df = n_{TOTAL} - k$

where $n_{TOTAL} = n_1 + n_2 + n_3 + $ etc.
and where k = the number of groups

Appendix E
t-Table

Critical t-values needed for rejection of the null hypothesis (see Fig. E.1)

T.J. Quirk et al., *Excel 2010 for Physical Sciences Statistics: A Guide to Solving Practical Problems*, DOI 10.1007/978-3-319-00630-7,
© Springer International Publishing Switzerland 2013

Fig. E.1 Critical t-values
Needed for Rejection of the
Null Hypothesis

sample size n	degrees of freedom df	critical t
10	9	2.262
11	10	2.228
12	11	2.201
13	12	2.179
14	13	2.160
15	14	2.145
16	15	2.131
17	16	2.120
18	17	2.110
19	18	2.101
20	19	2.093
21	20	2.086
22	21	2.080
23	22	2.074
24	23	2.069
25	24	2.064
26	25	2.060
27	26	2.056
28	27	2.052
29	28	2.048
30	29	2.045
31	30	2.042
32	31	2.040
33	32	2.037
34	33	2.035
35	34	2.032
36	35	2.030
37	36	2.028
38	37	2.026
39	38	2.024
40	39	2.023
infinity	infinity	1.960

Index

T.J. Quirk et al., *Excel 2010 for Physical Sciences Statistics: A Guide to Solving
Practical Problems*, DOI 10.1007/978-3-319-00630-7,
© Springer International Publishing Switzerland 2013